DEVELOPING BIOMASS POTENTIAL: TURNING HAZARDOUS FUELS INTO VALUABLE PRODUCTS

OVERSIGHT HEARING

BEFORE THE

SUBCOMMITTEE ON FORESTS AND FOREST HEALTH

OF THE

COMMITTEE ON RESOURCES
U.S. HOUSE OF REPRESENTATIVES

ONE HUNDRED EIGHTH CONGRESS

SECOND SESSION

Wednesday, June 23, 2004

Serial No. 108-99

Printed for the use of the Committee on Resources

Available via the World Wide Web: http://www.access.gpo.gov/congress/house

or

Committee address: http://resourcescommittee.house.gov

U.S. GOVERNMENT PRINTING OFFICE

94-533 PS   WASHINGTON : 2005

COMMITTEE ON RESOURCES

RICHARD W. POMBO, California, *Chairman*
NICK J. RAHALL II, West Virginia, *Ranking Democrat Member*

Don Young, Alaska
W.J. "Billy" Tauzin, Louisiana
Jim Saxton, New Jersey
Elton Gallegly, California
John J. Duncan, Jr., Tennessee
Wayne T. Gilchrest, Maryland
Ken Calvert, California
Scott McInnis, Colorado
Barbara Cubin, Wyoming
George Radanovich, California
Walter B. Jones, Jr., North Carolina
Chris Cannon, Utah
John E. Peterson, Pennsylvania
Jim Gibbons, Nevada,
   Vice Chairman
Mark E. Souder, Indiana
Greg Walden, Oregon
Thomas G. Tancredo, Colorado
J.D. Hayworth, Arizona
Tom Osborne, Nebraska
Jeff Flake, Arizona
Dennis R. Rehberg, Montana
Rick Renzi, Arizona
Tom Cole, Oklahoma
Stevan Pearce, New Mexico
Rob Bishop, Utah
Devin Nunes, California
Randy Neugebauer, Texas

Dale E. Kildee, Michigan
Eni F.H. Faleomavaega, American Samoa
Neil Abercrombie, Hawaii
Solomon P. Ortiz, Texas
Frank Pallone, Jr., New Jersey
Calvin M. Dooley, California
Donna M. Christensen, Virgin Islands
Ron Kind, Wisconsin
Jay Inslee, Washington
Grace F. Napolitano, California
Tom Udall, New Mexico
Mark Udall, Colorado
Aníbal Acevedo-Vilá, Puerto Rico
Brad Carson, Oklahoma
Raúl M. Grijalva, Arizona
Dennis A. Cardoza, California
Madeleine Z. Bordallo, Guam
Stephanie Herseth, South Dakota
George Miller, California
Edward J. Markey, Massachusetts
Rubén Hinojosa, Texas
Ciro D. Rodriguez, Texas
Joe Baca, California

Steven J. Ding, *Chief of Staff*
Lisa Pittman, *Chief Counsel*
James H. Zoia, *Democrat Staff Director*
Jeffrey P. Petrich, *Democrat Chief Counsel*

———

SUBCOMMITTEE ON FORESTS AND FOREST HEALTH

GREG WALDEN, Oregon, *Chairman*
JAY INSLEE, Washington, *Ranking Democrat Member*

John J. Duncan, Jr., Tennessee
Scott McInnis, Colorado
Walter B. Jones, Jr., North Carolina
John E. Peterson, Pennsylvania
Thomas G. Tancredo, Colorado
J.D. Hayworth, Arizona
Jeff Flake, Arizona
Rick Renzi, Arizona
Stevan Pearce, New Mexico
Richard W. Pombo, California, *ex officio*

Dale E. Kildee, Michigan
Tom Udall, New Mexico
Mark Udall, Colorado
Aníbal Acevedo-Vilá, Puerto Rico
Brad Carson, Oklahoma
Stephanie Herseth, South Dakota
VACANCY
VACANCY
Nick J. Rahall II, West Virginia, *ex officio*

———

CONTENTS

OVERSIGHT HEARING ON DEVELOPING BIO-MASS POTENTIAL: TURNING HAZARDOUS FUELS INTO VALUABLE PRODUCTS

Wednesday, June 23, 2004
U.S. House of Representatives
Subcommittee on Forests and Forest Health
Committee on Resources
Washington, D.C.

The Subcommittee met, pursuant to notice, at 2:00 p.m., in Room 1334, Longworth House Office Building, Hon. Greg Walden [Chairman of the Subcommittee] presiding.

Present: Representatives Walden, Rehberg, Renzi, Inslee, Tom Udall, and Herseth.

STATEMENT OF HON. GREG WALDEN, A REPRESENTATIVE IN CONGRESS FROM THE STATE OF OREGON

Mr. WALDEN. The Subcommittee on Forests and Forest Health will come to order. The Subcommittee is meeting today to hear testimony on "Developing Biomass Potential: Turning Hazardous Fuels into Valuable Products."

Under Committee Rule 4(g), the Chairman and Ranking Minority Member can make opening statements, and if any other Members have statements, they can be included in the hearing record under unanimous consent.

As hundreds of sawmills closed in recent years due to the shutdown of the Federal timber sale program, many lamented, including myself, at the loss of jobs and the debilitating impact to the economies of local communities. As early as the 1980s, some scientists and forest managers began warning of another impending crisis resulting from these closures. That would be the loss of infrastructure and markets for treating and funding the treatment of millions of acres of hazardous fuels. This admonition has already turned into reality, as many in this room know.

Many regions, as a consequence of losing local sawmills, also lost a well-trained and experienced workforce—equipment operators, loggers, truckers, and mill workers, not to mention the technology and infrastructure that these workers operated and managed. It has been lost. Now many communities have no alternative but to landfill or burn the timber and brush that they are removing in order to protect their communities, materials that could otherwise have been sold to help them offset the costs of treating local

(1)

forests. With 190 million acres of Federal lands at high risk of catastrophic fire, this is a very serious concern.

The primary purpose of today's hearing is to discuss issues surrounding the rebuilding of a viable infrastructure and to address a number of questions, such as what technologies and markets currently exist for the use of woody biomass and are they commercially viable? Have State or local governments promoted the use of biomass through subsidies, tax deductions or credits, loan guarantees, or other means, and how effective have they been? What technological, geographic, economic, or other obstacles exist for use and expansion of biomass? And what steps are Federal agencies taking to expand the use of biomass?

With the recent passage of the Healthy Forests Restoration Act and the vast amounts of woody material that are likely to be generated, the answers to these questions are even more urgent. Ultimately, the successful implementation of HFRA will require broad development of new industries and a rebuilding of traditional ones if our forests and communities are to remain viable and healthy.

To help us address this important issue, we are fortunate today to have a number of expert witnesses. With their insight, I hope we can begin to lay the groundwork for bipartisan Congressional action on biomass utilization.

[The prepared statement of Mr. Walden follows:]

Statement of The Honorable Greg Walden, a Representative in Congress from the State of Oregon

As hundreds of sawmills closed in recent years due to the shut down of the federal timber sale program, many lamented, including myself, at the loss of jobs and the debilitating impacts to the economies of local communities. As early as the 1980's, some scientists and forest managers began warning of another impending crisis resulting from these closures——the loss of infrastructure and markets for treating, and funding the treatment of, millions of acres of hazardous fuels. This admonition has already turned into reality. Many regions, as a consequence of losing local sawmills also lost a well-trained and experienced workforce; equipment operators, loggers, truckers and millworkers, not to mention the technology and infrastructure that these workers operated and managed. Now, many communities have no alternative but to landfill or burn the timber and brush that they are removing in order to protect their communities——materials that could otherwise have been sold to help them offset the costs of treating local forests. With 190 million acres of federal lands at high risk of catastrophic fire, this is a serious concern.

The primary purpose of today's hearing is to discuss issues surrounding the rebuilding of a viable infrastructure, and to address a number of questions, such as:

- What technologies and markets currently exist for the use of woody biomass and are they commercially viable?
- Have state or local governments promoted the use of biomass through subsidies, tax deductions or credits, loan guarantees, or other means, and how effective have they been?
- What technological, geographic, economic or other obstacles exist for use and expansion of biomass?
- What steps are federal agencies taking to expand the use of biomass?

With the recent passage of the Healthy Forests Restoration Act, and the vast amounts of woody material that are likely to be generated, the answers to these questions are even more urgent. Ultimately, the successful implementation of HFRA will require broad development of new industries and a rebuilding of traditional ones, if our forests and communities are to remain viable and healthy.

To help us address this important issue, we are fortunate today to have a number of expert witnesses. With their insight, I hope we can begin to the lay the groundwork for bipartisan Congressional action on biomass utilization.

Mr. WALDEN. I would like to introduce our witnesses today. On panel one, we have Chris Risbrudt, Director of the Forest Products Lab for the Forest Service, and if you want to come on up. I would normally turn to my Ranking Member when he arrives or someone on their side. They will have an opportunity to make an opening statement, but we will move ahead at this point.

So let me remind the witness that under our Committee Rules, you must limit your oral statement to 5 minutes, but your entire statement, of course, will appear in the record.

I now recognize you for your statement and we appreciate your coming here today to share your insights on your work in the lab, a very important part of our process. Thank you.

STATEMENT OF CHRIS RISBRUDT, DIRECTOR, FOREST PRODUCTS LABORATORY, FOREST SERVICE, U.S. DEPARTMENT OF AGRICULTURE

Dr. RISBRUDT. Thank you, Mr. Chairman, and thank you for the opportunity to meet with your committee today. I am Dr. Chris Risbrudt, Director of the Forest Products Laboratory in Madison, Wisconsin. The lab specializes in finding new and improved uses for wood and you have asked all of us today to speak about "Developing Biomass Potential: Turning Hazardous Fuels into Valuable Products."

The Healthy Forests Restoration Act signed into law last December by President Bush marks a clear and decisive change in direction to address the causes of catastrophic wildfires and insect and disease infestation by implementing hazardous fuel reduction projects in priority areas. I know you, Mr. Chairman, and the Subcommittee members recognize the scope of the threat to our forests and our communities.

The authorities in HFRA will help us to accomplish our mission. The one potential hurdle is the marketability of the millions of tons of woody biomass we will need to remove from the landscape. The lack of markets will lead to continued outlays of funds to remove material and then dispose of it.

We are here today to tell you about the new processes and products Forest Service researchers have been developing that will help overcome this hurdle. But before talking about some of these props I have got ready here, let me try to take the acres, condition classes, and stand densities that have been the focus of Congressional debate regarding the Healthy Forests Restoration Act and translate them into volumes of biomass and timber to give the Committee a greater appreciation of the immense stream of woody materials that will need to be disposed of after necessary thinning operations have taken place.

We have a report issued in April of 2003 entitled "A Strategic Assessment of Forest Biomass and Fuel Reduction Treatments in Western States" that was a joint effort involving a team of Forest Service researchers in cooperation with the Western Forestry Leadership Council, and that is a good source of information.

Let me state that healthy forests is not solely an issue in the West, but one for our entire country. But for our purposes today, I am concentrating somewhat on the West, where our greatest challenges lie.

The objective of the assessment was to characterize the amount of forest biomass that could potentially be removed to implement the objectives of the National Fire Plan. The assessment covers forests on both public and private ownerships and describes all standing tree volume, including stems, limbs, and tops.

First, the assessment noted that 15 Western States encompass almost a billion acres of land, of which 236 acres are forested. Slightly more than half of that forested area is classified as timberland. This acreage was further refined by fire regime condition classes, which is the measure of how much a forest has departed from its natural wildland fire condition. The scientists also employed plot data from 37,000 permanent FIA field plots, that is the Forest Inventory and Analysis plots, and they were summarized by forest type and ecoregion.

Let me make an important point. While removal of sub-merchantable seedlings and saplings is important to reduce ladder fuels, there is ample research that indicates that there is a range of stand condition where thinning only small-diameter material does little to reduce crown fire spread. There is also research indicating that a comprehensive treatment, that is one that removes some trees from all diameter classes, has a more significant effect on reducing fire risk than removing only small trees in many stand conditions, although that certainly helps.

The assessment provides several scenarios of the merchantable wood and biomass that could be produced. I will limit this discussion to two. Under one scenario, needed mechanical treatments done on 60 percent of fire regime Condition Class III lands would result in West-wide annual removals over 30 years of eight million bone dry tons of merchantable wood and 3.5 million tons of non-merchantable wood. The other scenarios where treatments would be done on both Condition Class I and II lands, and that results in 21 million bone dry tons of merchantable wood and 8.7 million tons of non-merchantable wood.

To put those figures into context, in 1999, the Western forestry industry processed about 28 million bone dry tons of roundwood for lumber and 2.2 million bone dry tons for pulpwood. We are currently estimating that we are removing 32 million tons of annual growing stock. So you can see that these two scenarios either represent one-third of the harvest or nearly equal the harvest of what we currently have.

Now, we have a number of efforts throughout the Forest Service, many being conducted jointly by research and development and State and private forestry that focus on three key areas for using large volumes of biomass—pulp and paper, energy and fuel, and engineered wood products and composites. But I would like to invite you, Mr. Chairman, and members of this committee to come out to the Forest Products Laboratory and see what we are working on. A member of this committee, Congressman Peterson, made the trip to Madison during one of our entrepreneurs tours, and I think he was excited about what he saw.

Thank you, Mr. Chairman, for your time, and I will be pleased to answer any questions you have about the assessment or about our programs at the Forest Products Lab.

Mr. WALDEN. Thank you. We appreciate your testimony today.

[The prepared statement of Dr. Risbrudt follows:]

Statement of Chris Risbrudt, Ph.D., Director,
USDA Forest Service, Forest Products Laboratory

Mr. Chairman, thank you for the opportunity to meet with your committee today. I am Dr. Chris Risbrudt, Director of the USDA Forest Service's Forest Products Laboratory in Madison, Wisconsin. The Lab specializes in finding new and improved uses for wood. You have asked me to speak about Developing Biomass Potential: Turning Hazardous Fuels into Valuable Products.

The Healthy Forests Restoration Act (HFRA) signed into law last December by President Bush marks a clear and decisive change in direction to address the causes of catastrophic wildfires and insect and disease infestations, by implementing hazardous fuel reduction projects in priority areas. This is a laudable and necessary goal.

I know you, Mr. Chairman, and the Subcommittee members recognize the scope of the threat to our forests and communities. The authorities in HFRA will help us accomplish our mission, but one potential hurdle is the marketability of the millions of tons of woody biomass we will need to remove from these landscapes. The lack of markets will lead to continued outlays of funds to remove material, and then to dispose of it. We are here today to tell you about the new processes and products Forest Service researchers have been developing that will help overcome this hurdle.

Before talking about that, I will try to take the acres, condition classes and stand densities that have been the focus of the Congressional debate regarding HFRA and translate them into volumes of biomass and timber to give you a greater appreciation of the immense stream of woody materials that will need to be disposed of after necessary thinning operations have taken place.

The April 2003 report entitled "A Strategic Assessment of Forest Biomass and Fuel Reduction Treatments in Western States" that was a joint effort involving a team of Forest Service researchers in cooperation with the Western Forestry Leadership Coalition is a good source. Let me state that healthy forests is not solely an issue for the West, but one for our entire country. But for purposes of this testimony today, I am concentrating somewhat on the West where our greatest challenges lie.

The objective of the assessment was to characterize on a regional scale the amount of forest biomass that could potentially be removed to implement the fuel reduction and ecosystem restoration objectives of the National Fire Plan for the Western United States. The assessment covers forests on both public and private ownerships and describes all standing tree volume including stems, limbs, and tops. The assessment includes analysis of treatment areas and potential removals, as well as the operational systems necessary to effect the treatments, the potential environmental impacts, and utilization opportunities for removed material.

First, the assessment found the 15 western states encompass almost 1 billion acres of land, of which 236 million acres are forested. Slightly more than half of the forested area (130 million acres) is classified as timberland according to the standard definition (i.e., capable of growing at least 20 cubic feet per acre per year and not reserved by law or administrative action from timber harvest). This acreage was further refined by Fire Regime Condition Class—which is a measure of how much a forest has departed from natural wild land fire conditions.

The scientists then estimated current forest conditions for areas needing hazardous fuel reduction treatments based on the combination of Forest Inventory and Analysis (FIA) data and a well accepted course-scale fire regime assessment. Plot data from 37,000 permanent FIA field plots were summarized by forest type and ecoregion. Computer modeling then applied selective removal prescriptions to that inventory using Stand Density Index (SDI) criteria. SDI is a long-established, science-based forest stocking guide that can be adapted to uneven-aged forests using data available from broad-scale inventories. This approach allowed for prescriptions across a wide range of ecosystems to reduce stand density to a healthy condition, determined in the assessment to be 30 percent of maximum SDI for any given stand. Trees assumed to be removed generally were small to mid-size trees. However, larger trees could also be removed if needed to reach an overall healthy condition for the forest and provide for regeneration of desired species.

This is important. While removal of sub-merchantable seedlings and saplings is important to reduce ladder fuels, there is ample research that indicates that there is a range of stand condition where thinning only small material does little to reduce crown fire spread. There is also research indicating that a comprehensive treatment, that is, one that removes some trees from all diameter classes, has a more significant effect on reducing fire risk than removing only small trees in many stand conditions. It also greatly improves the regeneration of desired species and reduces treatment costs to taxpayers.

The assessment excluded reserved forests and low-productivity forests and made reductions for operational limitations such as steep slopes, and sensitive sites. According to a global analysis, about 60 percent of the North American temperate forest is considered accessible (not reserved or high elevation and within 15 miles of major transportation infrastructure). A survey of National Forest land and resource management plans from 1995 also found that about 60 percent of the western National Forest timberland base is considered "suitable" for timber production operations (this is only 37 percent of the forestland base). The determination of "suitable" indicates that current forest operations technology would not produce irreversible damage to soil or water resources.

Applying the selective removal prescriptions to the identified inventory across the West, the assessment projected that the vast majority (86%) of the trees that could be removed would be less than 10 inches in diameter. There are nearly 2 billion trees in the 2-inch diameter class alone. While most of the trees that could be removed would be less than 10 inches, most of the associated volume would come from the 14 percent of the trees that are greater than 9 inches in diameter. In fact, under the assessment's projections, half of the volume would come from trees greater than 13 inches in diameter.

The assessment provides several scenarios of the merchantable wood and biomass that could be produced. I will limit this discussion to two: under one scenario, needed mechanical treatments done on 60 % of Fire Regime Condition Class III lands would result in West wide annual removals over 30 years of: 8 million bone dry tons (bdt) of merchantable wood and 3.4 million bdt of non merchantable wood, for a total of 11.4 million bdt. The other scenario is where treatments would be done on 60% of both Condition Class II and III lands. That could be project to result in West wide annual removals over 30 years of 21 million bdt of merchantable wood and 8.7 million bdt of non merchantable wood, for a total of 30 million bdt.

Put those figures into context. In 1999, the western forest industry processed about 28 million bdt of roundwood for lumber and 2.2 million bdt for pulpwood. Of the portion going to lumber mills, more than half the volume went as residues to pulp and particleboard mills. Current estimates indicate 32 million bdt of annual growing stock removals in the West are currently going to all products including medium-density fiberboard (MDF) plants, particle board plants, pulpwood and hog fuel. The scenario above involving only Condition Class III lands could represent about 36% of the current level of annual harvest in Western States (32 million bdt). Treatments of condition class II and III lands results in removals that are about 94% of the current level of annual harvest in Western States. Volume from thinning treatments could either replace current sources of raw material within the existing manufacturing infrastructure; or it could require private sector investment in new facilities.

The market price impacts from the fuels reduction program could range from practically nothing to very large. For example, a program that mechanically reduces fuels on Condition Class II and III forestlands and that simply added to current harvests could result in total region harvests of more than 60 million bdt and large aggregate price reductions. Price reductions arising from such a program might also negatively impact non-participating forestland owners through lower timber prices. A program that only addressed fuels on the Condition Class III lands but that replaced 8 million bdt of existing harvests would have much less aggregate price impact, although some local effects could be experienced.

The potential size of the manufacturing infrastructure needed to process material from fuel reduction treatments is large. Whether there would be expansion at existing facilities, restarted mills, or new construction would depend on many factors.

The economics of establishing a large number of processing facilities is highly uncertain. Attracting investment to new processing infrastructure involves analysis of long-term supply and market forecasts. Today's forest products markets are global and western production will have to compete with material from other wood producing regions. There are considerable challenges associated with establishing new processing plants in the West that go well beyond implementation of the fuel reduction treatments.

A complete analysis of the market effects as well as program costs will be conducted under a separate Joint Fire Science Program study, "A national study of the economic impacts of biomass removals to mitigate wildfire damages on federal, state, and private lands." This study seeks to evaluate market price and other economic effects of alternative scales of fuel reduction programs, with emphasis on Wildland-Urban-Interface zones. The study will also evaluate the differential effects of fuel reduction harvests that produce merchantable materials that substitute for or add to existing regional harvests.

So there is a challenge to find, grow or create markets and facility infrastructure sufficient to accommodate this volume of materials, much of which will come from small-diameter material for which there is not substantial market opportunities.

Congress did not ignore that pressing need in HFRA. Title II of the law provides authority to obtain information that will help overcome barriers to the production and use of biomass and help communities and businesses create economic opportunity. Three programs will help achieve these goals.

Section 201 of HFRA amends the Biomass Research and Development Act of 2000 to authorize research focus on overcoming barriers to the use of small diameter biomass. Many of the more than 120 proposals now being considered for funding under that Act by the Department of Agriculture and the Department of Energy relate to forestry and small diameter material. In all, some $22 million will be available this year. Forest Service Research and Development also has a comprehensive research program in the major areas of forest biomass assessment, management, harvesting, utilization, processing, and marketing.

Section 202 of HFRA, Rural Revitalization through Forestry, is aimed at helping communities and businesses create economic opportunity through the sustainable use of the nation's forest resources. While the key to this will be the actions of the private sector, the likelihood of success can be increased through the participation of State Foresters; Forest Service Technology Marketing specialists, such as at the Forest Products Lab; and federal and state economic development assistance agencies in collective efforts with local non-profit and for-profit businesses to build community-based forest enterprises. On-going efforts of the unit at the lab and S&PF resource specialists across the country provide this support.

Section 203 of HFRA authorizes grants to persons who own or operate a facility that uses biomass as a raw material for specific processes and products. The Forest Service has authority to provide grants for businesses, units of state and local government, non-governmental organizations (NGOs), and other entities with legal status. This Title expands authority to persons owning or operating facilities that use biomass as a raw material in producing energy, sensible heat, transportation fuels, and biobased products. Grants are limited to costs related to the purchase of biomass.

There are a number of efforts throughout the Forest Service, many being conducted jointly by R&D and State & Private Forestry that focus on three key areas for using large volumes of biomass: pulp and paper, energy and fuel, and engineered wood products and composites.

This hearing is focused on the third area. While I will discuss those programs at the Forest Products Laboratory which I know best, there are other important programs for forest products utilization in Forest Service Research and Development and State and Private Forestry which could focus on underutilized biomass.

The performance of new composite materials is determined primarily by the properties of the wood particles, the polymer binder, and the interfacial region that is established between the two distinct phases. Forest Service research at Pineville, Louisiana, is exploring the relationship between wood surface properties and interfacial characteristics, and addressing thermosetting and thermoplastic polymer systems to develop superior wood-based composite products.

Forest Service researchers in Blacksburg, Virginia, are developing and using expert systems and vision systems to support computer-aided and automated hardwood sawmill edging and trimming; developing a scanner/computer system to identify defects on rough lumber; supporting the development of a prototype vision system to automatically grade and upgrade rough lumber; developing products or better processes to improve the use of low-grade and small diameter hardwoods; and developing and evaluate automated production systems to grade pallet parts.

In Portland, Oregon, the program characterizes the forest resources and evaluates their uses by assessing the technical feasibility of producing primary and value-added wood products through empirical studies and simulation of western species. Projects such as establishing a database of western hemlock wood product recovery and lumber recovery from young-growth western hemlock and Sitka spruce in Alaska are the types of biomass work done by this program

The use of small diameter ponderosa pine that results from fuel reduction treatments is the focus of research in Flagstaff, AZ. This project is assessing the economic costs and benefits associated with different harvesting practices and regionally based utilization opportunities in fuel reduction treatments. This information will provide Federal land managers, contractors, and the public with an assessment of whether treatments can meet fuels reduction objectives at lower costs.

At the Forest Products Laboratory, we are working on a number of innovative engineered wood and composite products that could penetrate our nation's huge home building market.

For instance, this I-Beam, similar to those used extensively to support the floors in your home, is made out of tiny glulam beams sandwiched around a piece of oriented strandboard, or OSB. If you are not familiar with OSB, it is now used more commonly than plywood to sheath the homes in this country. Glulam beams are the large beams you'll find in many homes supporting the roof. Picture this one I'm holding here, only about 100 times larger.

The great thing about engineered wood products is that they can be made with virtually any fiber, including small-diameter timber. Builders love them because they are engineered and designed for a particular use. Because they are comprised of small pieces of fibers, they do not have knots and other flaws commonly found in solid wood. The strength in them is much more consistent. And they are much less likely to twist, bend, or warp.

Composites are another growth market that we are very excited about. Take a look at this shingle. It is made from recycled milk jugs and juniper. For those of you from the Southwest, you know some areas have an overabundance of juniper. It has taken over the landscape, crowding out other vital species and voraciously soaking up precious water. There is not much of a market for juniper—until now.

These shingles, which can be molded to look like Spanish tiles, cedar shakes, or whatever else you'd like, are just one example. They have a "class A" fire rating and an expected service life of 40 years. We are also working with a company in Mountainaire, New Mexico, to make signs out of juniper and plastic, such as this one that you might see on one of our National Forests. One of the biggest problems we've had with our signs is that porcupines love to eat them. However, they don't have an appetite for these. And they are much more resistant to a vandal's bullet than the old wooden ones. Although it sounds funny, these signs have proven to be very successful, and the little company in Ruidoso is now employing over 20 people, with plans to expand into other areas.

Another great idea our researchers have come up with is filtering contaminants from water with juniper. Filtering water is big business. These filters are very cheap to make, and very effective at removing contaminants such as acid mine waste, oils, pesticides, and agricultural and parking lot run-off. We also think they have great potential as erosion control mats. And you can use a variety of fibers. One possibility is using the slash from thinnings or the debris left after a fire to make erosion control mats to stabilize an area.

Energy is another high volume usage area. We are currently working with the DOE's National Renewable Energy Laboratory on a nationwide demonstration project using portable distributed energy systems. Distributed energy systems are decentralized energy production systems capable of grid connection. Basically, picture a large portable generator that you take with you to the woods, rather than bringing the woods to the generator.

The largest of the systems we will be demonstrating is 50 Kw in size, or about enough power to run about 10 residential homes. We feel that the results of these demonstration projects will then allow us to create a one-megawatt unit. A one-megawatt system would use about 12,000 tons of wood per year and produce enough electricity to power about 200 homes. And similar to what we've stated before, if you burn the unusable logs for power, sell the merchantable logs, and sell the power to the grid, you can actually make a profit while doing forest thinning. Other Forest Service research stations are developing management systems to ensure efficient and effective treatments; product development, utilization, and evaluation; and sustainability of the wood and bioenergy resource.

I could go on and on about our products, but I've got a lot of other people who are patiently waiting to tell their story. I'd like to invite everyone from this committee to come out to the Forest Products Laboratory to see what we're working on. Congressman Peterson, made the trip to Madison during one of our entrepreneur tours, and I think he was excited about what he saw.

For the past several months we have jointly hosted with Evergreen Magazine a series of tours for small business owners throughout the West to show them some of our small-diameter utilization technologies.

There are numerous specialty markets for small-diameter material such as post-and-rail, rustic furniture, firewood, animal bedding, and composts. Many of the witnesses today have success stories to share with you in these markets. We see opportunities both for large, volume driven businesses and for small, niche market driven businesses. Both sides will play a part in helping us solve the small-diameter problem.

Many people who would like to start a small forestry based business of some sort are doing it for the first time. They do not have the experience to pull things together like a business plan that will allow them to go to a bank and get a loan. That is where the Forest Service can help. We can help them decide what business makes sense for their given resources and market, and outline a specific course of action. Efforts like these are the key to restoring that lost infrastructure we talked about earlier.

Thank you, Mr. Chairman and committee members, for your time. I would be pleased to answer any questions you have about the assessment or our programs at the Forest Products Laboratory.

Mr. WALDEN. Let me go back to part of what you said about this research indicating a comprehensive treatment, that is one that removes some trees from all diameter classes, has a more significant effect on reducing fire risk than removing only small trees in many stand conditions—

Dr. RISBRUDT. Yes.

Mr. WALDEN.—because during the discussion on the Healthy Forests Restoration Act, we heard a lot about fuel treatment programs and spent a lot of time focused on ladder fuels and brushy understory. Can you be more specific as to the types of stands where the regiment you are talking about is more effective?

Dr. RISBRUDT. I think it is more effective in all stands that need that kind of treatment, which as you know is a very large acreage, something like 90 million acres or more. And so it depends on the density and composition of the stand.

It is just common sense when you think about it that the more fuel is there and how it is distributed, you get a bigger fire. And if you remove some of that fuel, it is less intense. I know there was some question about that, but we now have compiled a fair amount of research that shows that common sense does, in fact, apply in this situation.

Mr. WALDEN. To the extent to which you can share that with the Committee in addition to what you already have, that would be helpful.

You indicated you brought some, quote-unquote, props with you.

Dr. RISBRUDT. Yes.

Mr. WALDEN. Do you want to talk to us about that? Before you do, though, let me welcome the newest Member of Congress and the newest member of our committee and the newest member of our Subcommittee, Stephanie Herseth from South Dakota. We are delighted to have you join us on this panel and I know there are certainly issues in South Dakota revolving around forests and so we are glad to have you on board.

Dr. RISBRUDT. OK, thank you.

Mr. WALDEN. Let us go to the props.

Dr. RISBRUDT. All right.

Mr. WALDEN. What are you finding? What are you doing with all this wood product, because a lot of what we focus on is biomass for energy—

Dr. RISBRUDT. Yes.

Mr. WALDEN.—but clearly, we need to do more than that. So tell us what your lab has found and how we can be of help.

Dr. RISBRUDT. Here is an example of a laminated beam. The upper and lower portions of it are made like plywood with one exception. The grain always runs in the same direction. So this is about a dozen laminates glued together, top and bottom, and so you can make this out of pretty small material. Then the webbing in the middle that holds these two apart is made out of—this is a wafer board, and so you make that out of a very small diameter material and it makes a very good I-joist for floors. In fact, floors are stiffer with this material than they are when you build them out of—

Mr. WALDEN. We are going to grab those props and circulate them around the Subcommittee here, if that is all right.

Dr. RISBRUDT. Here is an example of a similar product with a plywood web in the middle.

Some of the things we are excited about are sort of non-traditional wood products. These are water filters made out of juniper, and as you are probably well aware, juniper is an invasive species across the West on our grazing lands, and so this is effective for taking—treating acid mine drainage. It takes the heavy metals out of acid mine drainage. And you also have to change the pH, but it takes oil, petroleum products out of parking lot runoff. We are taking pesticides out of cranberry bogs in Wisconsin and Massachusetts because those farmers have to treat for weeds and insects. It is also good for sediment, but you don't need—

Mr. WALDEN. Does the water taste like gin when you are done, or—

[Laughter.]

Dr. RISBRUDT. I haven't personally tried that, but I will do that.

Mr. WALDEN. Always a new use.

Dr. RISBRUDT. Yes.

[Laughter.]

Mr. WALDEN. We have a lot of juniper out there. What else do you have?

Dr. RISBRUDT. Here is an example of the water filter in the small scale of the current configuration, where we put it in a mesh, a wire mesh like this, and we do this for research purposes so right now it is kind of an expensive process, but we hope to go to this style where we just grind up the wood, particularly the bark, because of the chemical composition of the bark is very good at extracting ions and cations from water, and so we are hoping we can get to the stage where we just throw a bag of this material in the acid mine drainage, for example—

Mr. WALDEN. And so you can actually get rid of pesticides and oil products and acids with the juniper—

Dr. RISBRUDT. Yes. Now, the efficiency varies depending on the pollutant. But we are looking at treating the wood so it is even better at taking out certain materials.

Another product—and, of course, you can use extremely small-diameter material in these kinds of things where you are grinding them up. Another use we are looking at right now, this is made out of recycled milk jugs or plastics and this is pineflower—

Mr. WALDEN. It is what?

Dr. RISBRUDT. Pineflower.

Mr. WALDEN. OK.

Dr. RISBRUDT. Softwoods. We are also looking at doing this with juniper. Although the BLM has promised me a ton of juniper material, they haven't delivered yet for testing. We think it will make very durable siding. We have got samples of this for committee members.

We have got a company in Mountain Air, New Mexico, that is making signs out of ground up juniper and plastic. One of the problems the Forest Service and the BLM have with traditional plywood signs is the porcupines like to chew them up. They don't like plastic and juniper, so they are longer lasting.

So this is just an example of some of the products we are working on at the Forest Products Laboratory.

Mr. WALDEN. How economically feasible is the use of juniper, because, I mean, in my district, it is a noxious weed, and in fact a big one, but—

Dr. RISBRUDT. Well, for these specialty products, like water filters, that is a very large market. You can imagine just in the farming community where they have runoff from their fields with fertilizers, if we could develop these filters for that, the advantage there is once the filter gets loaded with pollutants, which is really fertilizer, you take it back out and throw it on top of the hill in the field and recycle it right into the field.

Mr. WALDEN. I will be darned.

Dr. RISBRUDT. So for these kinds of products, like siding, water filters, the material is the small cost of the final product.

Mr. WALDEN. Are you finding that there is enough of a market for that, for what you are coming up with for practical uses of these waste products? Do we need to do incentives, or will the market catch on? What makes this work?

Dr. RISBRUDT. Let me say there that if we are going to handle the volume of material that we know is available, you have to have a very large use for it and the largest use that I see that is feasible is turning it into energy. These are all valuable products. They will generate employment in communities. They will solve problems for the environment. But to handle the amount of volume we need, energy is the one we are going to have to develop and that takes incentives.

Mr. WALDEN. My time has expired. I will turn to the gentleman from New Mexico, Mr. Udall.

Mr. TOM UDALL OF NEW MEXICO. Thank you, Mr. Chairman.

In your testimony, you highlighted the importance of the biomass title of the Healthy Forests bill, at least the part of it I read here. Can you explain why the Forest Service did not request the authorized level of funding for the two grant programs authorized in the bill?

Dr. RISBRUDT. Yes. You will recall that the Healthy Forests Restoration Act was signed in December of just last year and so far, we are out of sync with the budgeting process and so we haven't had an opportunity to request those funds.

Mr. TOM UDALL OF NEW MEXICO. I thought that is when the budget process starts. That is right when you all start putting in—

Dr. RISBRUDT. We start three—

Mr. TOM UDALL OF NEW MEXICO.—requests and back and forth. There is still time to slip it in, isn't there? The budget doesn't come out until February, the President's budget, and then we are still looking at 2005 right now. You can change your mind right here and say that you are for it, that you want $5 million in each of those accounts.

[Laughter.]

Mr. WALDEN. That may be a question better for Mr. Ray to answer.

Mr. TOM UDALL OF NEW MEXICO. OK. Well, I just—I am trying to get him. Greg, I am trying to get him on the book here. He is the only one we have so far. I know he is a good guy—

[Laughter.]

Mr. TOM UDALL OF NEW MEXICO.—but the Department right now doesn't have a position on the two $5 million accounts, right, the two grant accounts?

Dr. RISBRUDT. They have not put anything as far as I know into the 2005 budget request.

Mr. TOM UDALL OF NEW MEXICO. OK.

Dr. RISBRUDT. They are still working on 2006.

Mr. TOM UDALL OF NEW MEXICO. Well, I hope you spread the word, at least from me, that I would like to see that we put funding into those accounts, and I am sure my colleague, the Chairman here, is going to be looking at that in the appropriations process.

A good deal of your testimony focused on the new wood products technology that utilizes small-diameter trees and fiber, and you passed some of these around to us up here. This is incredibly encouraging and points to a future where we can potentially leave behind the forest wars of the past. What type of outreach is the Forest Products Laboratory doing with industry and interest groups? How long do you think it will take for these products to be in the market? Is this technology also being developed globally or does it have the potential to give the American forest products industry a leg up in other countries?

Dr. RISBRUDT. We have started just this year something we call entrepreneurs tours, where we have contracted with Jim Peterson, who knows a lot of the business owners, mill owners, particularly the small and medium-sized companies, and we are inviting them to the Forest Products Laboratory for a day-and-a-half tour. So far, we have had two of those and they have proven to be, by the surveys and returns, they have proven to be very popular.

We are looking to do a series of them into the future. In fact, Congressman Peterson, who visited the lab several months ago, said he was going to round up his own entrepreneurs from Pennsylvania and bring them to the laboratory and we are looking forward to him doing that.

Beyond that, we have several newsletters, Newsline that we send out to thousands of people. We have a website. We send out our list of publications quarterly, any way we can think of to get the word out, bring people to the lab, let them learn about the products that are becoming available. We try our best and are anxious to try any new ideas that you may have for us.

As far as moving the products into commercial activity, it depends on, of course, the product, whether we can find entrepreneurs available, whether we can have grants available to help that process. I think we know what it takes to start businesses based on the technology at the Forest Products Laboratory. It takes an entrepreneur, it takes technical assistance, and it is not just the one visit, here, read this publication. It is a series of visits as they design and build their mill. It is business planning, market planning, and small grants to help them get started. If we can put those five things together, which we have at times in the past, we know how to get businesses started.

Mr. TOM UDALL OF NEW MEXICO. Great. Thank you. And the global side of this, I mean, are we way ahead on this or not?

Dr. RISBRUDT. Some of these products, I think we are, but the Scandinavian countries have large research programs and they are better integrated than we are between government, universities, and industry. And, in fact, with the globalization particularly of the pulp and paper side, we have—our local Wisconsin flagship company, paper company, was just bought by Stora-Enso, a Scandinavian-South American combination company. So it is difficult to tell. I like to think that we are in the lead on some of these things, but I would have a hard time justifying that if you look at the global scale.

Mr. TOM UDALL OF NEW MEXICO. Thank you very much. Thanks, Mr. Chairman.

Mr. WALDEN. Thank you for your questions.

I am informed, too, by the staff that apparently the Administration has received some grant applications under that title and they do intend to move some money to be able to deal with that once they receive the grants.

Dr. RISBRUDT. Yes. We had $22 million between USDA and DOE for those products, but they are not funded under that—it is to do that work—

Mr. WALDEN. Yes.

Dr. RISBRUDT.—but it isn't specifically funded for that program.

Mr. WALDEN. But under the $5 million you were talking about, apparently—

Mr. TOM UDALL OF NEW MEXICO. Two $5 million grant accounts.

Mr. WALDEN. My understanding is they are accepting applications and then will determine kind of how much they need to fund some of the appropriate ones.

Mr. TOM UDALL OF NEW MEXICO. That is good news.

Mr. WALDEN. I am sure your comments will be well heard, too.

Mr. TOM UDALL OF NEW MEXICO. That is good news, good news.

Mr. WALDEN. Yes.

Mr. TOM UDALL OF NEW MEXICO. I am sure Mr. Renzi, he will have a lot of applicants from his district, and Denny will, too, I am sure.

Mr. WALDEN. And I am now told next week, they will be announcing some of those grants. So if you just keep going here, they will have them out by this afternoon at this rate.

[Laughter.]

Mr. TOM UDALL OF NEW MEXICO. That would be great. Good work. Thank you.

Mr. WALDEN. Let us go now to the gentleman from Montana for 5 minutes.

Mr. REHBERG. Thank you, Mr. Chairman, and I want to point out that Mr. Peterson did, in fact, invite me on that tour and I wasn't able to make it, but he came back raving about what he had seen—

Dr. RISBRUDT. I am glad to hear it.

Mr. REHBERG.—and so congratulations to you on that. Are you the only center like that in the country?

Dr. RISBRUDT. The only Forest Service research station, yes.

Mr. REHBERG. OK, because, you know, you always find out there is more going on than you really believe. I took the time to go over to Sweden to look at what they had, with Bernie Sanders, and over there he told me, "Oh, we have been burning slash and are creating energy in Maine for years." I hadn't heard that and wish I had. It would have saved us a whole lot of time in Montana. How many small businesses do you think you deal with in a calendar year?

Dr. RISBRUDT. Oh, it is probably on the order of hundreds.

Mr. REHBERG. Hundreds?

Dr. RISBRUDT. Yes.

Mr. REHBERG. And you are a full-service operation? They come to you—somebody from Montana would come up with an idea—we have an idea called Timberwelt. I don't know if they are interested, but they create the great big huge beams and it is the same concept that I see traveling by. If they were to make contact with you, you would invite them out and you would talk about the cost-benefit—

Dr. RISBRUDT. Absolutely.

Mr. REHBERG.—and the kind of equipment and how—

Dr. RISBRUDT. Yes. In fact, we have a publication that we will make available to the Committee called "Small Diameter Success Stories" that lists, oh, I don't know, 20 or so small businesses that have started up using the technical resources of the lab and the technical assistance provided by State and private forestry.

Mr. REHBERG. That is always the frustration for those of us that—we know, like Mr. Udall says, there is a huge need in Montana for adding value.

Dr. RISBRUDT. Yes.

Mr. REHBERG. That is why we created the Innovation Center within the farm bill for agricultural products, and perhaps we should have tied it more closely to Healthy Forests. It is always frustrating to find that things are available that our people could be taking advantage of and are not aware. You have the staffing. You have the budget. It is the grants that you are having difficulty with?

Dr. RISBRUDT. Yes.

Mr. REHBERG. So you have the time to spend with people?

Dr. RISBRUDT. Yes.

Mr. REHBERG. Great. Thank you, Mr. Chairman.

Mr. WALDEN. Thank you. The Chair now recognizes the Ranking Member on the Subcommittee, Mr. Inslee, for questions or an opening statement.

Mr. INSLEE. I am going to yield. You have done such a great job so far, Mr. Chair.

Mr. WALDEN. Thank you. We will now go to Mr. Renzi.

Mr. RENZI. Thank you, Mr. Chairman.

Sir, I appreciate you coming today and your testimony. I had an opportunity to go down and visit an OSB plant down in Carthage, Texas. I had an opportunity to visit in my own district a biomass plant producing, I think it is three kilowatts of electricity onto the grid—megawatts, kilowatts—

Dr. RISBRUDT. Megawatts.

Mr. RENZI.—megawatts, and a good opportunity to go around and look at some of the other uses for biomass, including pellets for the furnaces, the saw logs that you buy in the grocery store, or the wood logs that you have in your fireplace.

During my recent trip to the OSB plant, I was told that one of the inconsistencies that commercial industry is looking at in the stewardship contracts that we are getting ready to let, one of them that is getting ready to come out in Arizona on the border between my district and Mr. Udall's, is 150,000 acres. It is 15,000 acres over 10 years. It is not enough to lure commercial industry into making an investment of $10, $15 million into an OSB plan or, I think you described it as a wafer board, similar product.

Dr. RISBRUDT. Yes.

Mr. RENZI. And yet we hear that there is an abundance of resources out there, particularly small diameter wood—

Dr. RISBRUDT. Yes.

Mr. RENZI.—and that there needs to be more done as far as the number of stewardship contracts that are put forward, the guarantees essentially that we need to put in place in order for commercial industry to have a reasonable return on their investment, to be able to lure them in. Can you expand on that a little bit, teach me a little bit about how it is that when you are letting a 150,000 acre contract people think it is a solution. It really isn't. We need to be up in the 300,000, 450,000 acre landscape which some people, of course, don't even want us in the woods—

Dr. RISBRUDT. Right.

Mr. RENZI.—so there is a little dichotomy there.

Dr. RISBRUDT. There are people in the two panels following me that might be able to give you a better answer, but I think it is not the acreage limitation, it is the timeframe. If you are going to put in $40 million into a plant, you have got to have—tell your banker you have a 20-year payback period and the stewardship contracts only run for 10 years. So a 150,000 acre contract may be enough for, I don't know, maybe one mill, but it is the timeframe that is the major restriction. You can't guarantee supply over 20 years, and that is where the bankers say, no deal.

Mr. RENZI. OK. So that is why we need to look at being able to layer those kind of stewardship contracts, one after another.

Dr. RISBRUDT. Yes. You need sufficient volume, certainly, on an annual basis. But it is the timeframe that you have to convince— the entrepreneurs and businessmen have to convince their banker that they will be in business long enough to pay off that 20-year loan.

Mr. RENZI. While I have got you, can you help me understand, is there a real issue with our being able to build biomass plants and them produce electricity and not be able to plug it into the grid? Is there some sort of a—

Dr. RISBRUDT. It depends upon the size of the plant.

Mr. RENZI. Yes.

Dr. RISBRUDT. When you have got a, say, a regular coal-fired plant, it is probably 500 megawatts or larger. Some of the big companies don't want to fool around with a smaller size and it is an irritation to them. I think that is kind of a cultural thing in the electrical industry that Congress may have to help them get over.

Mr. RENZI. Exactly. So if a three megawatt plant—

Dr. RISBRUDT. Is not very large.

Mr. RENZI.—which is not very large but which is average for this biomass industry, is that correct?

Dr. RISBRUDT. There are some people here who can tell you that better than I can.

Mr. RENZI. All right, maybe when I get a chance, you guys can help me learn that.

Mr. Chairman, thank you. Thanks so much.

Mr. WALDEN. Thank you. Thank you for coming today. We have all enjoyed seeing these products and look forward to learning more about these developments, and hopefully we will get an opportunity to go out to the lab. I would really enjoy that. Mr. Peterson also spent a lot of time with me on the Floor talking about how impressed he was, and Jim Peterson has done the same from Evergreen, a real advocate, so—

Dr. RISBRUDT. We would love to have you all.

Mr. WALDEN. Thank you. Thank you for being here today.

Mr. WALDEN. Now I would like to introduce our second panel of witnesses. On panel two, we have Mr. Bill Carlson, Vice President for Business Development, Wheelabrator Technology; Dr. David Tilotta, North Carolina State University; and Mr. Peter Johnston, Manager for Technology Development, Arizona Public Service.

Thank you, gentlemen, for coming out to Washington to share your views on biomass. We appreciate and look forward to your testimony. I would just remind you, too, under our Committee Rules, limit your comments, if you would, to 5 minutes. Your entire statement will be available to our members and in our official record.

Now I would like to recognize Mr. Carlson for his statement. Good afternoon.

STATEMENT OF WILLIAM H. CARLSON, VICE PRESIDENT, BUSINESS DEVELOPMENT, WHEELABRATOR TECHNOLOGIES, AND CHAIRMAN, USA BIOMASS POWER PRODUCERS ALLIANCE, REDDING, CALIFORNIA

Mr. CARLSON. Good afternoon. Mr. Chairman and Members, you have done yeoman work in recognizing the forest health crisis and in crafting solutions so that Federal agencies can address the crisis before all is lost to insects, disease, and fire. The new stewardship contracting authority, the National Fire Plan, and the HFRA have all passed before this Subcommittee.

With these new authorities and funding, the agencies are gearing up to improve forest health through prescribed fire and mechanical thinning. With 190 million acres at risk, even a five million acre per year program will take 40 years. So how do we mount the massive campaign we need to restore our Federal forests and rangelands in time with limited funds?

The agencies will find that the infrastructure of small log processing facilities and biomass power plants that would take and pay for the output of thinning does not exist outside Northern California. Without infrastructure, the cost will likely be $800 to $1,000 per acre, will run the agencies out of funding long before they meet their acreage goals. With infrastructure in place, the cost should fall to zero to $200 per acre, an amount we can afford.

We must create circumstances that allow infrastructure to be developed. Others today will discuss innovative ways to utilize the primarily small logs that are the product of these thinnings. All are needed, as well as two-by-fours and paper if we are to utilize the 250 million tons per year that may flow from a five million acre per year thinning program. Utilizing every last stick for higher valued uses, there will still be 40 percent of the material that will have no value other than as fuel.

This 100-plus million tons per year should go to biomass power plants to power 8,000 megawatts of needed domestic renewable energy. Getting the biomass power plants built to assist thinnings is no easy task, as it is moving against an economic current that has swept away nearly 40 percent of all U.S. biomass plants over the last 15 years. Low fossil fuel prices, utility contract buyouts, and an unusable Federal biomass tax credit have combined to doom many facilities.

One example, Mr. Chairman, from your own district, that of the Confederated Tribes of the Warm Springs, demonstrates the difficulty. The tribes have had a small, stable forest products industry, complete with sawmill and biomass power plant, utilizing logs from the tribes' own forest. The tribes have proposed to modernize and expand their sawmill and power plant to focus on the smaller average log size that will come from thinnings and to increase capacity to accept logs and fuel from adjacent Federal lands. The tribes' expanded facilities could become the utilization center for much of the thinning activity proposed for the East side of the Central Oregon Cascades.

Fortunately, the expansion decision coincided with a request for proposals for renewable power issued by PacifiCorp and the tribes submitted a proposal. The tribes and their many supporters are collectively holding their breath awaiting the outcome.

A renewable auction like this is typically dominated by wind power, with a lower delivered cost partly due to use of the same Section 45 tax credit that biomass plants are unable to use. Winning bids are typically only one to one-and-a-half cents per kilowatt hour above bulk system power, or about five to five-and-a-half cents per kilowatt hour.

If the Warm Springs bid is accepted, it will result in a low margin operation despite the advantages of having an existing plant and interconnect, a steam customer, and a low projected fuel cost. A new biomass operation on a new site would not stand a chance in this auction.

To allow competitive biomass power bids, we must utilize the Section 45 wind and biomass tax credit. Plants can currently qualify only by combusting closed-loop biomass, which is grown exclusively for burning, and something that has never been done commercially. We have long sought to change the definition to include the forest thinnings we use for both new and existing plants. Administration budgets, both Republican and Democratic, have included the requested changes. Several bipartisan bills and the pending H.R. 6 conference report include the changes, but none have been adopted.

Currently, the Section 45 changes sit in S. 1637, the Senate version of the FSC bill. The House version of the same bill, H.R. 4520, does not include an energy tax title but instead once again extends Section 45 without changes that would make it usable. If the House version prevails in conference, new biomass power infrastructure will not be built in support of thinning projects and existing plants will continue to close.

Three years ago, we left this Subcommittee with our "to-do" list that contained several needed policy changes to improve forest health. You have, to your credit, completed that to-do list with the exception of one item, the changes to the biomass tax credit just discussed. You who understand how biomass power facilities enhance and lower the cost of forest health activities must assist us in conference on the FSC bill by adopting the Senate energy tax provisions or by making the needed changes during any reauthorization of Section 45.

Our industry stands ready to invest tens of billions of dollars in new biomass power infrastructure in support of forest health activities, but only if we have economically viable projects, and that means a usable biomass tax credit. Thank you.

Mr. WALDEN. Thank you, Mr. Carlson. We will put together a letter to the Chairman of the Ways and Means Committee conveying your thoughts and our support for what you recommend.

[The prepared statement of Mr. Carlson follows:]

Statement of William H. Carlson, Vice President, Business Development, Wheelabrator Technologies, and Chairman, USA Biomass Power Producers Alliance

This Subcommittee has done yeoman work over the last several years in recognizing the forest health crisis on public lands in the U.S., and in crafting solutions so that federal land management agencies have the tools to begin to address the crisis before all is lost to insects, disease and fire. The new stewardship contracting authority, the National Fire Plan and the Healthy Forest Restoration Act (HFRA) are all pieces of the forest health solution puzzle that have passed before this Subcommittee.

So with these new authorities, and the funding that comes with them, the federal land management agencies are gearing up to begin a massive effort to improve forest health through a combination of prescribed fire and mechanical thinning. The effort needs to be both massive and sustained, as we have, by most accounts, 190 million acres at risk, and even a 5 million acre per year program will take nearly 40 years to do the job; and we do not have that amount of time when you consider we are losing 6-7 million acres per year to catastrophic fire alone. So the question becomes, how do we mount the massive campaign we need to reclaim and restore our federal forests and rangelands to health in the time we have left with the limited funds available?

In terms of the first tool, prescribed fire, I will leave to others the debate over potential escapes, air quality impacts and spotty results. I will confine my remarks to mechanical thinning, an area that I have participated in as a recipient and converter of the fuel fraction from such thinnings for nearly 20 years.

In ramping up mechanical thinning projects throughout the West from their traditional base in northern California, the land management agencies will quickly find that the infrastructure of small log processing facilities and biomass power plants that would take the output, and pay market rates for it, simply does not exist. Without the infrastructure, the cost of thinning will likely be $800-$1,000 per acre, a cost that will run the agencies quickly out of money long before they have met their allotted acres to be thinned for the year. By contrast, with infrastructure in place, the cost should fall to $0-$200 per acre range, an amount that could be covered by the allotted $760 million per year in the HFRA.

So the question to be asked and answered by the hearing today is how do we create a set of circumstances that will allow the infrastructure to be developed in support of the needed thinning so that costs are reduced and viable rural economic activity is created and sustained? Others on the panel today will discuss innovative ways to utilize the primarily small logs that are the product of these thinnings, and thus create additional value and lower net thinning cost. All of these products are needed, as well as a fair amount of 2 x 4s and paper, if we are to utilize the massive amount of material, perhaps 250 million tons per year, that will flow from a large scale thinning program of say 5 million acres per year.

Based on our experience, try as you may to utilize every last stick for higher valued uses, there will still be 40% or more of the material that will have no value other than as fuel. That 100+ million tons per year, will need to go to biomass power plants where it could power 8,000mw or more of needed domestic, clean, renewable energy.

But getting the biomass power plant built in support of large scale thinning is no easy task as it is moving against an economic current that has swept away nearly 40% of all biomass plants in the U.S. over the last decade. The combination of previously low fossil fuel prices, utility contract buyouts and an inability to qualify for an existing federal biomass tax credit has doomed many facilities.

It is in this environment that we are now looking to build new plants. To show the difficulty of infrastructure development, let me give you just one example, Mr. Chairman, from your own central Oregon district, that of the Confederated Tribes of the Warm Springs. For many years now, the Tribes have had a small, but stable, forest products industry, complete with sawmill and small biomass power plant, utilizing almost exclusively logs from the Tribe's own forests.

The B&B complex fires of last year, of which you are painfully aware, burned over 90,000 acres of prime federal timber and recreational lands, including touching on the reservation. The fires filled the air of central Oregon with smoke for weeks on end. This fire sensitized many in the area to the need for large scale thinning, both on and off the tribal lands. The Tribes have proposed to modernize and expand their sawmill to focus on this smaller average log size that will come from such thinning, and to increase capacity so as to accept logs from adjacent federal lands in support of thinning efforts. In addition, the Tribes propose to modernize and expand their power plant to accomplish the same purpose. With the proposed expansions in place, the Tribes' facilities could become the utilization center for much of the thinning activity proposed for the east side of the Cascades in central Oregon.

Fortunately for the Tribes, the decision to seek to expand the biomass power plant coincided with a request for proposals (RFP) for new renewable power issued by PacifiCorp, the Portland utility with which the Tribes are interconnected, and the Tribes submitted a proposal. A short list from that RFP has not yet been announced, and the Tribes and their many supporters are collectively holding their breath.

Typically a renewable auction such as this is dominated by wind power, which typically has a lower delivered cost and, for the last 12 years, has been able to use the same Section 45 Tax Credit that biomass plants have been unable to use. Wind bids typically hold winning bids to only 1- 1 1/2 cents/kwh above bulk system power, or about 5- 5.5 cents/kwh, and the winning bids in this auction will likely fall in that range as well. If the Warm Springs bid is accepted, it will make for a low margin operation, despite the advantages of having an existing plant and interconnect, a steam customer, waste fuel for a portion of their needs, and a low projected fuel cost for the remainder of their fuel. A completely new biomass operation on a new site would not stand a chance in this auction.

The missing piece of this puzzle that I referred to in my title, and the piece that would allow competitive biomass power bids, is the ability to utilize the Section 45 wind and biomass tax credit, which has been on the books since 1992 but unutilized by biomass power plants. This is because plants qualify only by combusting "closed loop" biomass, that which is grown exclusively for burning, and something that has never been done commercially. Waste fuels such as forest thinnings do not qualify. For over 5 years now, the USA Biomass Power Producers Alliance has sought to change the definition to include the waste fuels we and others use, and to make the credit available to existing plants as well as to new.

The last several Administration budgets, both Republican and Democratic, have included the requested changes; the changes have been the subject of several bipartisan stand alone bills; and the pending HR6 Conference report includes the changes. But none have made it over the goal line, plants continue to struggle and close, and the Warm Springs bid appears vulnerable.

Currently, an acceptable version of the Section 45 changes (except the in service date for new plants) sits in the Energy Tax Title of S1637, the Senate version of the Foreign Sales Corporation bill. The House version of the same bill, HR4520, which passed last Thursday, does not include the Energy Tax Title, but instead once again extends Section 45 without changes that would make it usable by such plants as the Warm Springs. Should the House version prevail on this point in Conference, the predictable result is that new biomass power infrastructure will not be built in support of thinning projects and existing plants will continue to close.

I last spoke to the Subcommittee on behalf of the USABPPA just over three years ago at a hearing on somewhat the same topic. At that time I left you with our "to do" list that contained several needed policy changes that would dramatically enhance forest health. In that three years you have, to your credit, completed that—to do—list with the exception of only one item. That item is the changes to the biomass tax credit just discussed. We call upon members of the Subcommittee, who understand how the existence of biomass power plants enhances and lowers the cost of forest health activities to assist us in Conference on the Foreign Sales Corp bill by adding the Senate Energy Tax Title or by making the needed changes during the reauthorization of the Section 45 wind and biomass tax credit.

Our industry stands ready to invest ten of billions of dollars in new biomass power infrastructure in support of forest health activities over the next two decades. But this will only happen if we have economically viable projects, and the key to that viability is clearly the existence of a useable biomass tax credit. Those needed changes are in our opinion the only additional order of business for Congress before large scale cost effective thinning and restoration can begin.

———

Mr. WALDEN. Now, I would like to recognize Dr. Tilotta for your statement. Thank you for being here, as well. You are welcome to give us your oral statement and your written statement will, of course, be part of our record. Good afternoon.

STATEMENT OF DAVID C. TILOTTA, DEPARTMENT OF WOOD AND PAPER SCIENCE, NORTH CAROLINA STATE UNIVERSITY, RALEIGH, NORTH CAROLINA

Dr. TILOTTA. Good afternoon. Thank you, Mr. Chairman and committee members, and thank you for providing me the opportunity to discuss the Coalition for Advanced Housing and Forest Products Research, or CAHFPR, as we call it. I am David Tilotta, President of CAHFPR, and also an associate professor of wood and paper science at North Carolina State University in Raleigh.

I don't know if you noticed, but many of the props that Dr. Risbrudt just showed from the Forest Products Lab were housing related and used biomass in terms of housing. As many of you know, housing construction is one of the largest uses of forest products in the United States. In fact, according to the National Association of Home Builders, the NAHB, the average American home is about 2,100 square feet and it contains just over 13,000 board feet of framing lumber and more than 6,200 square feet of sheeting, 2,300 feet of exterior siding, et cetera. Of course, the majority of the estimated 1.6 million new homes that will be built in America over the next year will use wood frame construction and a variety of wood engineered types of products like you just saw. So an increased demand for wood and related materials in new construction is expected to continue.

I am here today to talk about CAHFPR, and CAHFPR, I think, is a research success story for us and it is a good partnership lesson. CAHFPR's university research and development extension of the USDA Forest Service Forest Products Laboratory, the FPL located in Madison, and had its genesis back in 1998.

Before I continue, though, you may be wondering, why combine forest products research and housing? Well, I think the answer is pretty self-evident based on what we have already heard. It makes sense. It is one of the largest markets for forest products, as I said, and unquestionably, we all need better housing that is more affordable, durable, energy efficient, and disaster efficient. Of course, it is the largest investment, if not the largest investment most of us are going to make in our lifetime. So it only makes sense. It is only logical that the substance of that investment really be crafted and maintained with the same sort of good science and engineering principles that we use to get a spacecraft on Mars. So really, it is a good marriage, biomass utilization and housing.

Well, CAHFPR really is a new way of doing business in the arena of housing research. It maximizes the results and really minimizes the cost to the American taxpayer. The current situation in housing research in the United States is not a promising one. The national research and scientific capacity across the traditional sectors have been declining. For example, the Forest Service in general has lost about 50 percent of its scientists over the last 15 years. The Forest Products Laboratory in Madison, Wisconsin, which is really the only Federal wood research facility, has seen their ranks diminish from 700 in 1944 to about 240 or so today. Of course, their budgets remain flat, research funding has remained flat, and that translates into decreased dollars.

So the question becomes, how can we do more with less? Obviously, research funding is important if we are going to find advanced uses for some of these materials.

Well, the Coalition for Advanced Housing and Forest Products Research was formed in partnership with the Advanced Housing Research Center located in FPL, and many of you have been out there and seen them and taken a look at their house. The AHRC program was established in part as a response to the Partnership for Advanced Technology in Housing Program. But FPL really founded it so that they coordinate and streamline their wide-ranging housing research and development activities.

CAHFPR, our organization, actively identifies, coordinates, and executes research and development for housing and one of its major themes is to conduct R&D that responds to the construction, financing, and marketing of housing. In general, the organizational home of CAHFPR is at the FPL. In addition to providing universities with access to their scientists, the FPL also provides us with technical report reviews, webpage space, publication assistance, and dissemination services.

Let me see, here. The programs that we undertake are by invitation and the universities that we have in our masses are by invitation. Some of the research areas that we are looking at include things like termite-resistant materials, durability and natural disaster resistance, and both programs at the AHRC and CAHFPR are guided by an independently conducted national needs assessment, and that is very important for us, and that national needs assessment is being done by the NAHB research center located in Maryland. They are surveying the key stakeholders, including the academicians, the builders, the home owners, the insurers, and others to keep us honest, to make sure that we do research that is relevant, and that is very important.

Let me just sort of summarize this by saying that, really, CAHFPR is about partnerships, partnerships among and between the universities, the Federal Government, and industry. And industry is an important, and I don't have time to explain to you, tie-in there. And frankly, we believe that in order to advance the science and engineering aspects of the house and the entire American home experience, we really must work together.

Thank you, Mr. Chairman and committee members, for your time and I would be pleased to answer any questions that you have.

Mr. WALDEN. Thank you, Doctor. We appreciate your comments, as well, today.

[The prepared statement of Dr. Tilotta follows:]

Statement of Dr. David Tilotta, President, Coalition for Advanced Housing and Forest Products Research (CAHFPR), and Associate Professor, Wood and Paper Science, North Carolina State University

Mr. Chairman, thank you for providing me with the opportunity to discuss the Coalition for Advanced Housing and Forest Products Research, or CAHFPR, with your committee today. I am Dr. David Tilotta, President of CAHFPR and an Associate Professor of Wood and Paper Science at North Carolina State University in Raleigh, NC.

As many of you know, housing construction is one of the largest uses of forest products in the United States. According to the National Association of Home Builders (the NAHB), the average American home is about 2,100 ft2 and contains just over 13,000 board feet of framing lumber, more than 6,200 ft2 of sheathing, and around 2,300 ft2 of exterior siding. And of course, the majority of the estimated 1.6 million new homes that will be built in America over the next year will use wood-frame construction and a variety of wood-based products. Thus, the increased demand for wood and related materials in new construction is expected to continue, as well as a corresponding increased demand for wood products in the repair, remodeling, and renovation construction industries.

I am here today to discuss our Coalition for Advanced Housing and Forest Products Research. CAHFPR is a university research and development extension of the USDA Forest Service, Forest Products Laboratory located in Madison, Wisconsin (FPL). CAHFPR had its genesis in ca.1998, but before I continue, you may be wondering: Why combine forest products research and housing? Our answer is: because it makes sense! Housing is one of the largest markets for forest products. And, unquestionably, America needs more affordable, durable, energy efficient and disaster resistant housing that will only come from the latest technological advances. Additionally, housing is one of the largest, if not the largest, investments that an individual makes in his or her lifetime. It is only logical that the substance of that investment be crafted and maintained with the same good science and engineering principles that allowed our nation to successfully land a spacecraft on Mars.

Simply, CAHFPR is a new way of doing business in the arena of housing research that maximizes the results and impact while minimizing the cost to the American taxpayer. Let me elaborate. The current situation in housing research in the United States is a rather dismal one: the national research and scientific capacity across all the traditional sectors (i.e., industry, university and government) have been declining, and international competition has been increasing. As an example, the Forest Service in general has lost, and not replaced, almost 50% of its scientists over the last 15 years. And more specifically, the Forest Products Laboratory in Madison, Wisconsin, the only Federal wood research facility, has seen their employee ranks diminish from 700 in 1944 to around 240 today. And of course, their total budget has remained approximately flat in recent years, which explicitly means that the amount of funding available for research has declined.

The trends in research funding, obviously important to university research, are not likely to dramatically change any time soon. So, we must ask the question: "How can we do more with less?"

The Coalition for Advanced Housing and Forest Products Research was formed in partnership with the Advanced Housing Research Center, or the AHRC, at the FPL to respond directly to the current diminishing research support from Washington, DC. Although the AHRC was established, in part, as a response to the Partnership for Advancing Technology in Housing (or PATH) program, FPL founded it so that they could coordinate and streamline their wide-ranging housing research and development activities.

CAHFPR actively identifies, coordinates and executes research and development (R&D) for housing, and one of its major themes is to conduct R&D that responds to the construction, financing, and marketing of housing. Universities are invited to participate in CAHFPR based on their expertise in a given programmatic area, e.g., termite resistant materials, durability and natural disaster resistance, etc. Research areas and programs that the AHRC and CAHFPR undertake are guided by an independently conducted national needs assessment. This year, that assessment is being done by the NAHB Research Center located in Maryland. They are surveying key stakeholders (e.g., academicians, builders, homeowners, insurers, and others) to identify and rank the most important research needs across the country. In general, CAHFPR research progress is assessed for quality and program relevance to ensure progress and efficiency via an external working group comprised of representatives from academia, industry and the government.

As I mentioned earlier, CAHFPR is a university extension of the FPL, and the linkage to them is to their AHRC. We provide expertise that is complimentary to the scientists and engineers that are members of their center. This direct linkage provides for a degree of systematic synergy and augmentation of effort that otherwise wouldn't be possible.

Organizationally, the virtual "home" of the coalition is at the FPL. In addition to providing the universities with access to their scientists, they also provide them with technical report reviews, web page space, publication assistance, and dissemination services. The dissemination of the research results to industry is important, and the vital feedback from them is handled by the FPL through their Residential Moisture Management Network technology transfer group. The Network is a government/industry association. As an aside, the FPL has also formed a separate group, the Federal Agency Housing Partnership, that coordinates housing research and technology transfer within and among various Federal agencies.

The AHRC and CAHFPR believe that
- long-term partnerships based on programmatic needs is the most efficient and direct means of impacting the most significant problems, and
- research should be crosscutting and integrated in order to optimize efficiency and maximize the benefit to the American public.

Current and future research areas of CAHFPR and the AHRC include:
- Moisture management and indoor air quality
- Improved use of traditional wood products
- Recycled and engineered wood composites
- Energy, sound, and environmental efficiency
- Natural disaster resistance
- Improved durability of finishes and sealants
- Better utilization of small diameter timbers and "junk" species

A long-term research union of the FPL, industry, the member universities of CAHFPR, and other affiliated government agencies has numerous advantages:
- The university researchers are allowed to network with some of the world's best scientists and engineers at the FPL, thereby increasing productivity
- The government has the benefit of working with, and training, the current generation of students, who will of course be the next generation of problem solvers and/or consumers
- The government does not have to duplicate research capacity that is present at the universities
- Together, the government and the universities can work on complex problems that necessarily may span years or decades to unravel
- Fast response to problems because the universities are "on call"

Partnerships are sometimes difficult to initiate and sustain, and CAHFPR has been no different. Since it's formation, some of the difficulties that we've had include:
- University administrators who adopt the "Why can't we have it all?" approach
- Partners who merely give lip service to partnerships
- University researchers who want funding without accountability
- Difficulty with existing laws that hamper the concept of "long-term" relationships with the Federal Government

But perseverance is prevailing. CAHFPR is currently stable with six universities as members. In FY05, we will add additional ones to our roster having expertise in fire-related housing issues and the utilization of small diameter timbers.

Finally, let me end this Statement be referencing a report that was published in 2002 by the National Research Council (the NRC, Cubbage, et. al, National Academy Press, Washington, DC, 2002) about four years after the beginnings of

CAHFPR. The USDA Forest Service requested that the NRC examine the national capacity for forestry research. And out of 11 "Recommendations," the following four are particularly relevant with respect to CAHFPR, and point out that we're on the right track:
- "The Forest Service should substantially strengthen its research workforce over the next five years to address current and impending shortfalls..."
- "As part of the increase in research personnel capacity and resources, the Forest Service should enhance cooperative relations with forestry schools and colleges."
- "The USDA, together with universities, should develop means to more effectively communicate existing and new knowledge to users, managers, and planners..."
- "Centers of excellence in forestry should be established and administered by USDA. These programs and awarded projects should (1) support interdisciplinary and interorganizational activities, (2) focus on increasing minority student participation in education and research, (3) clearly justify how new forestry-research approaches and capacity will be enhanced, and (4) undergo initial and periodic review."

As I said before, CAHFPR is about partnerships—partnerships with and among universities, the Federal Government (the USDA Forest Service Forest Products Laboratory), and industry. We believe that in order to advance the science and engineering aspects of the house and the "American Home," we must work together.

Thank you, Mr. Chairman and committee members, for your time. I would be pleased to answer any questions you have about CAHFPR.

———

Mr. WALDEN. Now I would like to recognize Dr. Johnston for your testimony, 5 minutes oral, and your written testimony will be in the record. Good afternoon, and welcome.

STATEMENT OF PETER JOHNSTON, MANAGER FOR TECHNOLOGY DEVELOPMENT, ARIZONA PUBLIC SERVICE COMPANY, PHOENIX, ARIZONA

Dr. JOHNSTON. Good afternoon, Mr. Chairman. My name is Peter Johnston and I am the Manager for Technology Development for Arizona Public Service. I appreciate the opportunity to talk to you today.

Arizona Public Service, or APS, is the largest electric utility serving the State of Arizona and we are currently working toward generating up to 1.1 percent of our electricity from renewable resources. To date, we have installed over five megawatts of solar photovoltaic generating plant. We have biomass, landfill gas, and wind projects. And we are exploring geothermal resource in Arizona and also the use of human and animal waste.

The biomass plant that we have running, one of the three-megawatt plants in the town of Eager, came online in February, and we have a second three-megawatt plant under construction destined for the town of Snowflake.

The biomass resource we see in the State of Arizona we believe has a sustainable capacity of somewhere between 200 and 500 megawatts. The largest renewable resource is, of course, solar, but biomass is a very significant resource that we would like to use. So we would like to extend more of these power plants and there are a couple of issues that have arisen that give us concern. One is the cost of the actual fuel we burn in the plants and one is the long-term availability of those fuel sources.

A biomass power plant of the size we are looking at, three megawatts, will probably never be competitive with a natural gas or a coal plant. That is not the issue for APS at the moment. The issue is that we need these plants to be competitive with other forms of renewable resource generation.

The plant at Eager, the cost of the fuel is just under $10 a ton and the cost of the energy from that plant is between seven and eight cents a kilowatt hour. That is more than twice the cost of energy from a conventional plant. If the price of the fuel goes up to around $30 or $40 a ton, then the cost of energy will go up to over ten cents a kilowatt hour and that would not be competitive with other forms of renewable generation and could limit the rate of expansion of our biomass activities.

We recognize that a great cost savings can be achieved if a biomass plant is installed in the same location as a wood product operator similar to these sort of products you have seen this afternoon, we can have a symbiotic type of relationship whereby the biomass plant actually takes the waste material from the wood processor, uses that as fuel in the electric generating part of the plant, and the generating plant can actually provide process heat to the wood operator. So it is a very neat relationship and, in fact, we are pursuing that type of relationship with several small diameter wood fabricators in the State.

To date, the progress of those partnerships has been hindered by the uncertainty of the fuel supply for the wood product fabricator, the OSB manufacturer. So one of the issues we have today is that the stewardship contracts should really look at the long term, making the fuel available on a long-term basis. I believe I was told that 5 years is the norm, 10 years is possible, and I would stress that 10 years is an absolute minimum for some of these operators to get financing and actually come into existence.

Until those fabricators do have viable operations, the biomass plants can continue if there is some form of fuel subsidy program. There used to be one in the State of Arizona. That would just give some moderation or continuity or certainty to the price of the fuel that we could use in the biomass power plants.

APS is committed to developing clean, renewable energy sources today that will fuel tomorrow's economy. We see biomass as being one of—a viable component of the renewable energy portfolio. Fortunately today, APS is able to pay a small premium for the energy coming from biomass power plants, something that may not continue into the future as larger-scale plants are developed. But we would like to continue to pursue the biomass energy ventures and we look forward to the cooperation and support from all parties to make those ventures successful. Thank you.

Mr. WALDEN. Thank you for your comments. We appreciate them.

[The prepared statement of Dr. Johnston follows:]

Statement of Dr Peter Johnston, Manager for Technology Development, Arizona Public Service Company

Good afternoon Mr. Chairman. My name is Peter Johnston and I am the Manager for Technology Development for Arizona Public Service (APS), an electric utility based in Phoenix Arizona, and I appreciate the opportunity to testify today.

APS is the largest electric utility serving the state of Arizona and is currently working to generate 1.1% of its retail electricity from renewable resources by the year 2007 in accordance with the state's Environmental Portfolio Standard (EPS). My department has been tasked with achieving that goal and has already completed a number of electricity generation projects from renewable energy sources located in the state.

As part of our program, we have completed an assessment of the renewable resources in the state and determined that after our most abundant resource, solar energy, and potentially, wind energy, biomass is the largest resource that APS can use towards meeting the EPS requirement.

We estimate that the ponderosa pine forests and pinon and juniper populated woodlands can support an electric generating capacity of between 250 and 500 MW. To that end APS has already funded the construction of a 3 MW biomass power plant that came on-line in February 2004 in Eagar, in Eastern Arizona. This plant is now generating electricity from the forest residues from the Wildland Urban Interface initiatives associated with the Apache-Sitgreaves forest. We are currently in the process of constructing a second 3 MW biomass plant that should be completed in early 2005 in Snowflake AZ and we are actively seeking additional plants to add to our biomass portfolio.

The cost of electricity generated from these plants is very dependent on the proximity of the fuel source to the plant and we have identified 12 locations in the state where electric generating facilities of between 3 MW and 40 MW could be sensibly built with good access to forest residues and electrical transmission infrastructure. Eagar and Snowflake are two of those identified locations.

The cost of electricity from these plants will be approximately 2 to 3 times that of electricity generated from a more conventional 300 MW or so natural gas fueled plant. This is due to a number of reasons relating to the relative size of the biomass units, however, the cost of the biomass fuel is a significant operating expense for a biomass power plant. A typical cost of biomass fuel is in the range of $10 to $40 per ton. In the case of the Eagar plant, for example, the fuel cost is just under $10 per ton and constitutes 20% of the annual Operating and Maintenance costs of the plant. The resulting cost of energy from Eagar is 7.68 c/kWh. If the fuel cost increased towards the top limit of $40 per ton, the energy cost would increase to more than 12 c/kWh. Naturally, APS would prefer the lower cost of energy to make the plant more competitive with other renewable resource opportunities. The cost of fuel can be reduced if a third party operation, which creates added value from the wood feedstock, can be sited at the power plant site and associated with the power plant operation.

Such an operation could be a fabricator of glulam boards for example, which makes construction boards from small diameter forest thinning material. The product from this plant has its own market value and the waste material from the plant's operation can be disposed of as the fuel feedstock for the electric generating plant. Additionally, process heat required for the glulam operation can be provided by the power plant thus saving the glulam plant operator the expense of constructing and operating a heating system. A symbiosis of this nature can actually result in a negative fuel cost for the power plant and the combination of the two operations can make the disposal of waste material leaving the forest profitable and ultimately enhance the economic development of the predominantly rural areas where the plants would be located.

APS is pursuing operating partnerships with several small diameter wood product companies in order to minimize the cost of electricity production from existing and future biomass power plants. The progress of these activities has been hindered by the uncertainty of a feedstock supply to the wood product companies. In order to finance their operations a feedstock availability of at least ten years is generally required and although recently awarded stewardship contracts can provide some level of that certainty, no such contracts have been released in Arizona to date. It is imperative for the successful deployment of small diameter wood product operations and additional biomass power plants in Arizona that the owners of these operations know that they will have access to a feedstock/fuel supply for at least ten years into the future.

Until such wood product companies are able to commence operations, residues from forest health operations can be collected and hauled to biomass power plants for conversion into electricity. As indicated above, the cost of electricity generated from a biomass plant is sensitive to the cost of fuel. Since fuel can be produced in areas not necessarily close to the power plants that exist or being planned today a Fuel Subsidy for hauling companies, as originally made available in HR2646, can have a decisive influence on the success or failure of the power plant as such subsidies can be used to moderate the cost of fuel hauled to the plant.

APS would like to continue to expand the number of biomass power plants in Arizona. Not only will they assist APS in meeting their EPS requirements but they will also provide a means of disposing of residues resulting from the healthy forest initiatives and provide economic development opportunities in the state. An added benefit resulting from the presence of these power plants will be a means of disposing of the more than 8,000 tons per month of urban green waste material that is currently disposed of in landfills in Arizona. Burning the waste in a controlled manner in a boiler will be an improvement over filling landfills. However, in order to achieve this expansion, the cost of the renewable electricity generated from these plants will have to be competitive with other renewably sourced electricity. The association or partnership of a biomass power plant with a value added operation will greatly facilitate this expansion.

We recognize that there are many factors that can influence the development and success of biomass to energy power projects some of which can be instigated at the federal level. APS would therefore encourage the following actions:

1. Congress should continue to provide funding to the Forest Service programs as proposed in the current Forest Health bill.
2. Encourage the National Forest Service to continue long-term NEPA preparations and issue Forest Stewardship contracts as soon as possible.
3. Continue to support community involvement in the issuance and approval of Forest Stewardship contracts.
4. Until such Stewardship contracts are forthcoming re-activate the availability of a Fuel Subsidy program.

APS is committed to developing clean, renewable energy sources today that will fuel tomorrow's economy. We see biomass as a viable component of our renewable energy portfolio. We also recognize that virtually all renewable energy projects require some form of financial subsidy to make their economics work. Fortunately APS is able to pay a premium for the electricity produced from renewable sources through the Environmental Portfolio Standard program. However, we are also cognizant of the fact that our program is open to the scrutiny of our regulators and our customers who expect our expenditures to be made prudently.

Turning Hazardous Fuels into Valuable Products has the potential to provide new job opportunities, local economic development and the creation of healthy forests for everyone's benefit. APS will continue to pursue renewable biomass energy ventures and looks forward to the cooperation and support from all parties to make those ventures successful.

This concludes my prepared testimony. Once again I appreciate the opportunity to speak before you today and will be glad to answer any questions you or the subcommittee might have.

———

Mr. WALDEN. I want to follow up on one of the points you made about the need for long-term guaranteed supply, as in the stewardship contract, because I have heard that from people who have the biomass fuel plants as well as others looking at these markets.

Dr. JOHNSTON. Yes.

Mr. WALDEN. How critical is that to getting capital to invest in biomass facilities and what kind of time line do they need?

Dr. JOHNSTON. It can be very critical as we found with some of the companies that we are trying to venture with. With APS, it has not been an issue. We are financing these on an expensed nature. We are not financing these, long-term financing for these power plants. We are simply expensing them.

But for people like Louisiana Pacific, a company that makes oriented strandboard, obtaining financing is important to the setting up of a new business. I don't know their time line, sir. We were supposed to have a combined plant in Arizona sorted out by January of this year. Unfortunately, 6 months have slipped. That is really—the three-megawatt power plant we are building in Snowflake was never intended for Snowflake. It was intended for the Flagstaff area. But since the financing didn't come through for the land appropriation there, we have located the plant into Snowflake.

Mr. WALDEN. Where is Snowflake?

Dr. JOHNSTON. Snowflake is a little to the northeast of Phoenix.

Mr. WALDEN. Does Snowflake ever get snow? That was the question.

[Laughter.]

Mr. TOM UDALL OF NEW MEXICO. It used to before global warming.

Mr. WALDEN. Before global warming. Yes, all right.

[Laughter.]

Mr. WALDEN. Don't steal the Ranking Member's line there, Mr. Udall.

[Laughter.]

Mr. WALDEN. Mr. Carlson, could you explain more about just technically what we need to get changed in this biomass closed-loop provision so that we can make that provision really workable for the projects and all you outlined?

Mr. CARLSON. Certainly, Mr. Chairman. That particular provision, as it is written, has been on the books actually since 1992 and it has worked extremely well for the wind power industry, which has been able to expand dramatically over that 12 years. But because it was drawn so narrowly to be, as I mentioned, just closed-loop biomass, which is material grown specifically for burning, it has been unusable and there has never been a dime collected by any biomass power producer.

Basically, the definition needs to be opened up, and there has been over the last several years a definition developed for basically three categories of fuel. There is fuel related to forestry operations, such as thinnings, mill waste materials. There is the category of agricultural fuels, such as orchard prunings, grape prunings, orchard removals, nut shells, that sort of thing. And then there is the urban wood category, which includes things like old used pallets and two-by-fours and that sort of thing, and everyone has become comfortable with that change in the definition.

So the definition has become almost portable. It moves around between bills or between the President's budget and everybody is comfortable with that, but it just never seems to make it over the goal line. There never seems to be one of these bills that actually gets implemented. It is set for a couple of years now in the energy bill, but we just don't have an energy bill.

Mr. WALDEN. All right. I appreciate that.

Dr. Tilotta, could you speak further about the research being done on recycled and engineered wood composites, and specifically, how close are we to economical and large-scale production?

Dr. TILOTTA. To the engineered wood products?

Mr. WALDEN. Right.

Dr. TILOTTA. That is not something that I can address.

Mr. WALDEN. Really? OK. Is anyone else on the panel able to address that issue?

[No response.]

Mr. WALDEN. OK. All right. I don't have any further questions.

Mr. Inslee?

Mr. INSLEE. Mr. Carlson and Dr. Johnston, maybe you could give us some thoughts about how to evaluate the impact of these tax benefits to these nascent industries. I mean, it is the same kind of issue whether it is solar or wind or biomass. It is the same kind of issue. How do we evaluate their effectiveness? Some people—and I am a big believer in them, so I am an advocate. But some critics have suggested, well, no, these are either going to happen or they are not on their own economics and these things are really of marginal utility in actually spurring investment. Give us your assessment of how we judge that issue, how you would judge that issue. Why don't we start with Mr. Carlson.

Mr. CARLSON. Thank you. Certainly. I think the best judge of that, quite honestly, is one that I just mentioned that we have a 12-year track record in, and that was that in 1992, the wind and biomass tax credit in Section 45 of the tax code was established. At that time, there was a fairly vibrant biomass industry based on the early what are called PURPA contracts, Public Utility Regulatory Policy Act that was passed in 1978, and there was a relatively infant wind industry because it was more risky from the standpoint of investment because they only generated when the wind blows, as opposed to biomass plants that generate all the time on a 24/7 basis.

Since that time, where the wind industry could easily use this credit, they have probably increased their capacity in the U.S. by probably tenfold over that period. Biomass plants, conversely, with the expiration of some of those contracts and the fact they were above market, has continued downhill to where about 40 percent of the plants that were online at that time are now gone.

So I think that gives you an indication. If you go back to some of Dr. Johnston's comments of a moment ago, talking about these plants having to be at market by some point, we can't ask the utilities to provide the above-market needs of the biomass plant longterm. I mean, they are in the business of generating electricity and selling it to customers and giving the customers the best deal. If we could have a usable biomass tax credit, the biomass plants would be able to put bids into those utilities that are very, very close to market and that is the real difference that you will see.

Quite honestly, in my opinion, it is the difference between—in the situation we face today and the focus of this committee, which is forest health, if we have a usable biomass tax credit for both new and existing facilities, you will see all the biomass facilities built that are necessary to support the thinning activities as the government ramps it up. It is the difference between seeing those plants built and basically piling and burning that material and the woods for the foreseeable future.

Dr. JOHNSTON. I agree with Mr. Carlson. I think a great example is the wind industry today. Their 1.8 cent per kilowatt hour tax credit is tied up in the energy bill and it is my understanding that a whole bunch of wind projects have stalled in 2004 as a result of that bill not coming through yet.

The amount of the subsidy is going to be important. One-point-eight cents for wind does make wind energy in a number of cases competitive with more conventional forms of generation. One-point-eight cents supplied to solar energy would have a very negligible effect. The cost of solar energy today is around 30 cents a kilowatt hour, so you would be looking for a tax credit of 25 cents or more before that had a significant impact on utility use of solar.

Biomass, depending on the size of the plant, 1.8 cents may be insufficient to encourage utilities to buy biomass energy. That may need to be a little more than that. So the size of the plant that APS is working on is not a commercial size plant, really. It is simply based on the limited budget we have to meet the portfolio standard.

So any, I guess, tax credit that you could give to a biomass generation facility would assist it to sell energy to utilities that do have portfolio standards to meet as opposed to just simply buying the energy for commercial use compared to a natural gas plant, for instance.

Mr. INSLEE. I know solar has, although still above market, has experienced reductions with increasing units sold. Basically, there is a curve, I think.

Dr. JOHNSTON. Yes.

Mr. INSLEE. I am told that every time it goes up by a factor of ten, the price comes down by ten or 20 percent or some factor there. Is biomass in the same situation?

Dr. JOHNSTON. No, sir. I don't think so. Solar energy is still a very new industry and it is a declining cost industry as those technologies are perfected. The biomass industry is very much, I think, a mature industry, although there are some new technologies coming out to use biomass in the form of either gasification or pyrolysis and we are working with the National Renewable Energy Lab, looking at the formation of bio oil that is almost equivalent to petroleum crude oil but its feedstock is a biomass material.

Mr. INSLEE. Interesting. Thank you.

Mr. WALDEN. Thank you, Mr. Inslee.

Now I go to the gentleman from Montana for 5 minutes.

Mr. REHBERG. Thank you, Mr. Chairman.

Mr. Carlson, I was listening, but I am not sure I heard the answer, and that is the reason why it wasn't in the House bill is we were asleep at the switch? We just didn't get it in? Or is there some opposition to doing it that you sense? You had mentioned the President and some others. I just want to know in my mind, is it just we weren't paying attention? Should we have gotten it in?

Mr. CARLSON. To my knowledge, there is no real opposition—there is no opposition to the provision that I know of. We have worked with some of the environmental groups, for instance, relative to the definition and everyone seems to be reasonably comfortable with the definition, the expanded definition. It was more a case of Chairman Thomas simply not wanting an energy tax title as part of the Foreign Sales Corporation bill but needing to reauthorize an existing whole bunch of tax credits that had expired. It really just got lumped in, and if it goes as it is written today, it will be reauthorized like it has been four or five times in the past without change.

Mr. REHBERG. OK.

Mr. CARLSON. We are trying to avoid that. To be quite honest, our coalition that is working on this is getting so thin now because of the plants that have closed that this may be our last shot at correcting this at this time.

Mr. REHBERG. OK. Dr. Johnston, you had mentioned that you essentially need something besides the electric generation or the biomass, some peripheral industry. Have you worked out those numbers specifically? The 1.8, you say, may help. How much would a peripheral, in your mind, small business, a Timberwelt type of a facility or whatever, some of the examples that we were sent, how much does that equate into a cost savings per kilowatt?

Dr. JOHNSTON. I can give you an example based on Eager, where I say we are paying $10 a ton for the fuel, and that constitutes around one cent per kilowatt hour in the cost of energy. If the fuel was zero cost instead of 7.6 cents a kilowatt hour, we would be down at 6.5 cents a kilowatt hour. So I don't think we could get a negative fuel cost. We might do it, depending on the industry we are working with, but we are looking at maybe reducing the cost of our energy by one or two cents.

Mr. REHBERG. Is the industry that you essentially work with, and maybe this is a question for the rest of you, as well, is yours essentially reliant upon access to Federal properties or do you have tribal and private properties, as well?

Dr. JOHNSTON. We have all of the above. We favor private properties if we can.

Mr. REHBERG. And why is that?

Dr. JOHNSTON. Just simplicity of contracts.

Mr. REHBERG. Access.

Dr. JOHNSTON. Yes, sir.

Mr. REHBERG. So our access laws become an impediment to your being able to have an additional facility or access to enough of the kind of product you need to create the energy on—

Dr. JOHNSTON. They have not been so far.

Mr. REHBERG. Mr. Carlson?

Mr. CARLSON. We have about 20 years' experience in Northern California doing this type of thing on an integrated basis, where the infrastructure does exist in all cases. The small wood forest products industry is there, cardboard plants, paper mills, within a reasonable haul distance, and we find that it makes all the difference in the world relative to thinning whether it be on private lands or on public lands in that most of the cases in Northern California, the land owner actually gets a return rather than making a payment. The bids will actually come in that they will pay him a couple hundred dollars an acre for access to accomplish the thinning.

That is one of the advantages, quite honestly, that Montana still has in developing biomass plants is that you still have a lot of that infrastructure. You still have a paper mill in Missoula and a hardboard plant in Missoula and several small family owned sawmills that make it far more economic in a place like Montana to complete that puzzle with a biomass plant than it would, say, in Arizona, where most of that infrastructure is now gone.

Mr. REHBERG. And then again, that is where the 10-year versus the 20-year ability to amortize—

Mr. CARLSON. Well, that is right, because that just become cost of power. I mean, you can build it on a 10-year basis, but you have raised the cost of power now a penny or a penny and a half per kilowatt hour versus a 20-year agreement.

Mr. REHBERG. Would it be safe to say that the infrastructure that is in place, let us use one of my small mills in Eureka, is it safe to say that they have access to the power grid because of the mill being there, that the facility is in place, or would there be technological changes necessary to build a grid, access to the grid so we can get it in the power system.

Mr. CARLSON. Well, let us use that example, and I don't know the specifics of the mill in Eureka, but if it is a typical sawmill, it will have an electrical—

Mr. REHBERG. It is pretty strong because it is still open—

Mr. CARLSON. Yes.

Mr. REHBERG.—amazingly so.

Mr. CARLSON. It has to be. It will have electrical load probably of three to five megawatts of power coming into the area. Well, you could easily—say it is five megawatts. You could easily then build a ten megawatt power plant on that same grid because you would displace the five megawatts the mill used and then turn around and send the five megawatts back out on the same system. So typically, if the plant doesn't get too large—if you are talking about a 50-megawatt plant, for instance, you may need a higher voltage than is available in Eureka, where it may only be a 12 KV system, as an example. But certainly something twice the size of the largest existing industry that is there could be accommodated.

Mr. REHBERG. Could I ask a follow-up question?

Mr. WALDEN. Certainly.

Mr. REHBERG. Would it then be cost effective for them to do it the 10 year, or still impossible, they would need 20 years to pay—

Mr. CARLSON. It is not impossible, like I say, it is just that it just raises the cost of the power. I mean, we were fortunate to get, quite honestly, the 10-year stewardship contract authority, so no one is truly arguing with that at the moment. Would 20 years give you a more cost-effective product to sell to someone like Arizona Public Service? Certainly, it would.

Mr. REHBERG. Thank you, Mr. Chairman.

Mr. WALDEN. Thank you for your comments.

We now turn to the gentleman from Arizona, Mr. Renzi, for 5 minutes.

Mr. RENZI. Thank you, Mr. Chairman.

Mr. Carlson, I appreciate your optimism, particularly in your statement when you talk about the idea of possibly having five million acres per year on a large-scale thinning program, which I think you said would produce 250 million tons per year. How do you gain that optimism? Is that just what you see that we don't?

[Laughter.]

Mr. CARLSON. Well, I guess I base that on the fact that certainly the people in this room, your intention is to actually solve the problem, and the problem is that we have 190 million acres of overstocked Federal land that needs to be thinned, and we all know that we are losing six to seven million acres a year by fire alone, not to mention what is being destroyed by insects and disease.

So if we truly intend to solve the problem in our lifetime, so to speak, we need a five million acre a year—and the Forest Service talks about programs reaching that level, of five million acres a year. Now, a large fraction of that in their case will be prescribed fire rather than mechanical thinning.

But even to go back to the contract you are talking about at the Apache-Sitgreaves of 150,000 acres over 10 years, that 15,000 acres a year will produce enough fuel—the fuel fraction of it alone is probably enough for 20 or 25 megawatts. So it is not an insignificant contract that we are talking about here. In fact, the Salt River Project currently has a request for proposal on the street for a ten megawatt biomass plant that may well be fueled by the residual of those thinnings in that particular instance.

Mr. RENZI. We are talking about, if there are ten stewardship contracts out there right now and each of them are conducting 15,000, we are talking about 150,000 acres a year. So we would have to—I mean, we are talking about exponentially having to let many, many more contracts and layer them and spread out these years so that wouldn't—

Mr. CARLSON. No, that is certainly true. One of the concepts that we have kicked around for many years, and I have been working on this concept for a long time, like close to 15 years now, was that every ranger district on every national forest in the West ought to have a biomass plant of 20 or 25 megawatts. If they did, that ranger district then could thin their entire range district with, again, using all the other forest products uses that you can over a period of 20 years.

Mr. RENZI. No, I want to get there with you. I have counted nine or ten stewardship contracts I think we are working on right now, Mr. Chairman, with your leadership, so we are taking 150,000 total acres, and if we are looking at five million acres, we have got quite a lot more work to do. But I am with you on the optimism and we will hopefully get there together. It is a beginning. It is a great first step.

Dr. Johnston, I appreciate you coming in from Arizona and thank you for your leadership. I had a chance to visit the plant in Eager and tour it. I was interested by your comments in following up on Mr. Inslee's line of questioning in that biomass is a mature industry and that the costs and driving down the costs are a little bit harder now. You also talked about the threat of costs going up because of the increase in material costs, is that correct?

Dr. JOHNSTON. Yes, sir, and that was primarily due to the location, I guess, of the forest thinnings and the transportation of those thinnings to the biomass plant. We would try and keep a plant and the source of fuel within a 50-mile radius. We go above 50 miles, the cost of transportation gets excessively high.

Mr. RENZI. And thus the need for more stewardship contracts that will have that ability to overlap and cut down on the transportation, I imagine.

Dr. JOHNSTON. Yes, sir.

Mr. RENZI. My colleague from Montana talked a little bit about the grid. What is this issue that we talked about, I am trying to learn about up in Eager, Arizona, as it relates to discrimination on the grid, of being able to access the grid? Is it location, as Mr. Rehberg was talking about? Is it because you are dealing with such a small quantity that—

Dr. JOHNSTON. I don't think it was access to the grid. I think it was access to the land, was it not, that your colleague was addressing.

Mr. RENZI. No, I was specifically talking about during my visit in Eager, Arizona, the fact that because the plant out there, the biomass plant is only producing three megawatts, that it is seen as a drop in the bucket and that—

Dr. JOHNSTON. Well, three megawatts is actually sufficient to feed the whole Town of Eagcr, so I guess it is all relative.

Mr. RENZI. Well, they go to bed early at night.

[Laughter.]

Dr. JOHNSTON. No, we burn wood there 24 hours a day, sir.

[Laughter.]

Dr. JOHNSTON. Compared to APS's overall capacity, we have over 4,500 megawatts of capacity that we own.

Mr. RENZI. I appreciate that.

Dr. JOHNSTON. So yes, it is a very small amount compared to that.

Mr. RENZI. Let me finish with one question to Professor Tilotta, hopefully with some hope here and optimism. I was looking at helping an operation locate to Flagstaff that was going to be involved in laminated wood, taking a small diameter wood, x-raying it with a computer, cutting it the best way the grains need to go in order to laminate it. I was told that there is great profit in that product, and since we have seen so many products come around the room today, can you give me one piece of hope?

Is laminated wood, the profit margin, is that the product that allows this kind of good profit margin to exist, or are we moving in that direction compared with maybe not as much profit margin in some of the sawdust logs? In some of the tours that I have done and some of the industries that I have seen, whether dealing with pellets or sawdust logs, the margins are so thin that there is not a whole lot of grand hope. Can you give me some optimism to finish up here?

Dr. TILOTTA. No, I agree. I think that application clearly is a high-profit one. In terms of if you look at the spectrum of things that you can do, all the way from wood products to paper and pulp sorts of applications, that is where I would go and put my money. But if you go all the way to the other end, perhaps, and I am going to look a little bit in the future, one application might be simply to use that—extract sort of the chemicals and the energy, if you will, directly from that. For example, extract chemicals that can be used for building blocks for other sorts of applications as well as ethanol and those sorts of things. So it is on the horizon, but I think it is coming.

Mr. RENZI. Thank you, Mr. Chairman.

Mr. WALDEN. Thank you, Mr. Renzi.

I want to thank the panel. Your testimony has been most enlightening and helpful as we look to better techniques to use biomass and what we can do in the Congress to assist this industry. Thank you for being here and thanks for sharing your comments. We may get back to you with other questions.

I would like to invite up panel three. We have Mr. Masood Akhtar, President, Center for Technology Transfer; Mr. Tom Coston, Fuels for Schools Coordinator, Bitter Root Resource Conservation and Development Area; Ms. Lynn Jungwirth, Executive Director, Watershed Research and Training Center; and Mr. Jason Drew, Director of the Nevada Tahoe Conservation District, National Association of Conservation Districts.

I would remind our witnesses you have 5 minutes for oral statements. Your written comments will be put in the record. We are told we are going to have some votes soon, at about 3:30. We may be able to get through the panel and ask a few questions, so please go right ahead.

Let us start with Mr. Akhtar, your statement, please, sir. Thank you and welcome. Would you make sure your microphone is turned on there.

STATEMENT OF MASOOD AKHTAR, PRESIDENT, CENTER FOR TECHNOLOGY, INC.

Mr. AKHTAR. Thank you, Mr. Chairman. Today, my testimony will focus on a main outcome of the Forest Products Industry Technology Alliance, which we all know is part of Agenda 2020, which is a program, a partnership between the governments, the forest products industry, and academia to develop technologies capable of increasing energy efficiency, reducing environmental impact, and improving the industry economics.

The Alliance highlighted the need to establish a proper technology development organization that will work very closely with the industry user, the funding agencies, the State, and the Federal regulatory agencies, and the businesses developing and marketing the technology. Once a technology has been identified, it must be presented properly to the industry.

In addition to establishing the technology development organization, the concept of a biorefinery has to be further explored. This concept has the potential for doubling profits to the industry by producing value-added products from biomass onsite while the industry can continue making their conventional paper products. We need your support for these initiatives, a dedicated technology deployment organization and biorefineries.

Today, I will share with you a Wisconsin technology deployment model that would easily be replicated throughout the U.S. with some modification, depending upon each State's need to improve the competitiveness of our U.S. forest products industry. The Wisconsin Department of Administration through its program called Focus on Energy created a nonprofit organization in 2002 called the Center for Technology Transfer—we call it CTT—which is basically a technology deployment organization. The mission of CTT is to improve the competitiveness of Wisconsin industry clusters, including the forest products industry.

As you know, Wisconsin is still the number one paper-producing State in the nation. In order to expedite the commercialization and implementation of federally funded technologies, the organization presents a package to the industry which includes the technology, funding for technology demonstration, energy and tax incentives for a limited period to early adapters.

Governor Jim Doyle of Wisconsin and his administration are fully committed to help the forest products industry in Wisconsin. His administration has made important progress in reducing permitting times and reforming the way State agencies do business. They have negotiated and signed legislation creating the Green Tier program within the Department of Natural Resources, DNR. This voluntary program encourages greater environmental performance by recognizing companies with superior environmental performance. More importantly, it rewards them with the increasing flexibility and less regulatory risk when trying new technologies.

He also signed legislation to create a sales tax exemption on energy used in manufacturing. Energy is a major cost, as you know, for manufacturers in the paper industry, the largest consumer of energy in Wisconsin. The bill will help companies stay competitive, stay in business, and stay in Wisconsin. This shows our State's serious commitment to providing a business-friendly environment.

Governor Jim Doyle has also announced his strong support for developing Wisconsin's renewable energy resources. The farm bill's energy title provides Federal resources that complement the Focus on Energy program—obviously the CTT is a part of that—and other efforts. With your support, the farm bill energy title programs can reach their full potential for our State and our nation. Looking into the future, we hope you will support an aggressive expansion of these programs in the next farm bill to meet the many challenges we face. With your support, Wisconsin can continue to lead the Nation in developing renewable resources and supporting rural communities.

The Governor also created a Governor's Council on Forestry in Wisconsin. In the recent meeting on June 17, the Council discussed priority issues of Wisconsin woodlands, which are owned by 260,000 people and others. They identified invasive species as the top priority.

This document provides further details on some of the issues I have outlined here. I hope the information provided here will help you make the decisions that are needed to improve the competitiveness of industry. My expertise is mostly in the area of technology transfer, which is critical for this committee, so I will be taking some questions at the end.

Mr. WALDEN. Thank you. Thank you very much for sharing your comments.

[The prepared statement of Mr. Akhtar follows:]

Statement of Masood Akhtar, President, Center for Technology Transfer, Inc.

My testimony will focus on a main outcome of the Forest Products Industry Technology Alliance. (This "Agenda 2020" program is a partnership between governments, the forest products industry and academia to develop technologies capable of increasing energy efficiency, reducing environmental impacts, and improving industry economics).

The Alliance highlighted the need to establish a proper Technology Deployment Organization that will work very closely with the industry user, the funding agencies, the state and the federal regulatory agencies, and the business developing and marketing the technology. Once a technology has been identified, it must be presented properly to the industry. Understanding the conservative nature of this industry, it is very critical that the new technology be presented in one-on-one meetings with industry representatives at three levels: 1) Company executive (focus on profit potential), 2) Research executive (focus on technical merit), and 3) Operating personnel (focus on how risk can be minimized). In addition to establishing the Technology Deployment organization, the concept of a "Biorefinery" has to be further explored. This concept has the potential for doubling profits to the industry by producing value-added products from biomass on site, while the industry can continue making their conventional paper products. We need your support for these initiatives, the dedicated Technology Deployment Organization and Biorefineries.

Today I will share with you a Wisconsin Technology Deployment model that could easily be replicated throughout the US, with some modifications depending upon each state's need, to improve the competitiveness of our U.S. Forest Products Industry. The Wisconsin Department of Administration through its Focus on Energy Program, created a non-profit organization in 2002 called the Center for Technology Transfer (CTT), which is basically a Technology Deployment Organization. The mission of CTT is to improve the competitiveness of Wisconsin industry clusters, including the Forest Products Industry. As you know, Wisconsin is still the number one paper-producing state in the nation. In order to expedite the commercialization and implementation of federally-funded technologies, this organization presents a package to the industry which includes the technology, funding for technology demonstration, energy and tax incentives for a limited period to early adopters, etc.

Governor Jim Doyle of Wisconsin and his administration are fully committed to help the forest products industry in Wisconsin. His administration has made important progress in reducing permitting times and reforming the way state agencies do business. They have negotiated and signed legislation creating the "Green Tier" program within the Wisconsin Department of Natural Resources (DNR). This voluntary program encourages greater environmental performance by recognizing companies with superior environmental performance. More importantly, it rewards them with increased flexibility and less regulatory risk when trying new technologies. He also signed legislation to create a sales tax exemption on energy used in manufacturing. Energy is a major cost for manufacturers in the paper industry, the largest consumer of energy in Wisconsin. The bill will help companies stay competitive, stay in business, and stay in Wisconsin. This shows our state's serious commitment to providing a business-friendly environment.

Governor Jim Doyle has announced his strong support for developing Wisconsin's renewable energy resources. The Farm Bill's Energy Title provides federal resources that complement Wisconsin's Focus on Energy program (CTT is part of this Program) and other efforts. With your support, the Farm Bill Energy Title programs can reach their full potential for our state and our nation. Looking to the future, we hope you will support an aggressive expansion of these programs in the next Farm Bill to meet the many challenges we face. With your support, Wisconsin can continue to lead the nation in developing renewable resources and supporting rural communities.

The Governor also created a Governor's Council on Forestry in Wisconsin. In a recent meeting on June 17, 2004, the Council discussed priority issues for Wisconsin Woodlands which are owned by 260,000 people and others (about 61% of the acreage total). They identified invasive species as the top priority. John Cutis, Wisconsin's foremost expert on vegetation in the state, found in the 1920-30's a healthy number of native species. Recent inventories of those same plots found a dramatic decrease in the variety of native species, and an alarming increase of non-native species that are destroying our natural ecosystems. This has the potential to do substantial damage to our wood-using industry and the economy of Wisconsin. Federal programs for forest health are important if we want to retain the vital forestry industry base in Wisconsin.

This document provides further details on some of the issues I have outlined here. I hope the information provided here will help you make the decisions that are needed to improve the competitiveness of our Forest Products Industry.

CTT Model: A Technology Deployment Organization in Wisconsin

The Center for Technology Transfer Inc. (CTT) is fueling Wisconsin's long-term economic growth by helping state researchers and entrepreneurs bring new energy- and cost-saving technologies to market. Based in Madison, WI, CTT is a one-stop-shop for commercializing new technologies. The private, nonprofit corporation helps established and early stage companies statewide move new technologies along the path from discovery to successful implementation in the marketplace, where they make state businesses more competitive and retain and create jobs.

Services to established companies

Identify and bring industry-specific technologies. In the competitive global environment, retaining jobs has become as critical as creating them. Implementing new technologies to reduce production costs is one way to retain high paying manufacturing jobs in Wisconsin. By understanding the nature of key Wisconsin industry clusters, CTT can help facilitate this process by:
- Arranging one-on-one-meetings with key industry managers to identify their industry-specific technology needs

- Searching for available technologies, particularly those that are funded by federal agencies like the U.S. Department of Energy, which are near commercialization or have been commercialized elsewhere
- Conducting initial technology, business, and financial due-diligence in cooperation with an established vendor with industry credibility
- Presenting appropriate technologies to senior managers of potential customers along with the vendor and developer.

The now prescreened technology can be evaluated by industry to determine if the output and investment's rate of return are acceptable. A plant demonstration on a pilot scale is the likely next step.

Provide funding for technology demonstration. To minimize risk to the industry, CTT can provide up to $250,000 to fund technology demonstrations. These funds are provided to the technology developer either in the form of equity, loans (secured or unsecured) or a combination of both. CTT can also leverage its funds by bringing in additional funds if needed.

Identify business and policy issues that are barriers to implementing technology. During one-on-one meetings with an industry cluster's representatives, CTT may identify barriers to implementing technologies. In response, CTT may prepare unbiased research reports comparing Wisconsin business incentives and policy issues with those of neighboring states and abroad. CTT will then arrange personal meetings with industry leaders, state agencies and others to develop strategies to overcome these barriers. CTT has completed such a review for the Forest Products cluster, and additional reviews are in progress.

Provide education and training. CTT provides industry-specific technology and training through interaction with trade organizations, universities, and technical colleges. Our current focus is to bring available technologies to users through existing conferences, trade shows, and the like.

Services to early stage start-up companies

CTT's assistance typically falls into one or more of the following areas:

Project Funding, including secured low-interest loans, unsecured loans, bridge loans for repayment or conversion to equity, and equity investment.

Business Mentoring, including offering a database of service providers, conducting due-diligence reviews, providing business planning advice and assistance, serving on boards of directors and boards of advisors, and advising on strategic negotiation with potential business partners.

Grant Assistance, including identifying available grants, providing personalized grant writing training, writing and reviewing grant applications, providing grant administration assistance, conducting technical due-diligence reviews, helping obtain letters of support and collaboration for grant applications, offering matching funds, and arranging bridge loans to sustain clients between federal grant phases.

Intellectual Property Assistance, including developing intellectual property strategies and assisting in the patenting and licensing of inventions, particularly for non-university inventors.

Wisconsin Forest Products Industry Challenges

The following list of perceived challenges facing the Wisconsin forest products industry are based primarily on five individual discussions between a task force and representatives from three paper mills and two sawmills.

FORESTRY BUSINESSES IN GENERAL

International:

Policy

- Foreign governments absorb some of the workers compensation through government paid benefits like health insurance.
- Foreign entities cannot own public utility power generating facilities in the US.
- Global environmental regulations vary—businesses need a level playing field.
- International Trade Barriers affect market access.
- Disparity between tariffs levied on imports into our domestic market and those imposed by other countries need correcting.

Business

- All facilities surveyed face international competition. The forest products industry is rapidly being integrated into the global economy.
- Effective global marketing strategies are needed for the forest products industry in Wisconsin.

- Increased competition from China in furniture and paperboard, Canada in softwood products, Europe and others in pulp and paperboard, and other forest products from Chile, Scandinavia, New Zealand, and Russia.
- Exchange rates affect multinational company decisions on where to make facility/capacity investments.

Education

- Organize a national seminar broadening what Bob Seavey, Dept. of Wood & Paper Science, University of Minnesota set up. "Manufacturing Strategies for Profitability in the 21st Century: Surviving Globalization" March 7, 2003.
- Develop a compilation of successful strategies used by companies to find niche markets and other methods to cope with globalization.

National:

- Present tax laws do not favor investment.
- There is a lack of available fiber from national forests in Wisconsin. (It has been suggested that different national forests are able to provide significantly different quantities of wood for use by industry.)
- It would be helpful to overhaul the Fair Labor Standards Act (FLSA) so that it allows an employer to give incentives to all employees without having to endure onerous calculations to adjust for overtime considerations each time you choose to award bonuses.
- Need to work with Department of Commerce to help solve policy issues.
- The Byrd amendment regarding softwood lumber needs to be reviewed.
- Need a study of present infrastructure to determine what adjustments can be made to improve it. (Need to encourage innovative research to support infrastructure improvements.)
- Need approval of categorical exclusions for small timber sales on federal lands.

State:

PERMITS:

Policy

- Permits are a major problem due to complexity and long time delays. Some companies believe there is no real way to cooperate with State of Wisconsin permitting agencies. Companies are reluctant to make even small changes because of the permit process.
  - It takes too long to get things done. Supposedly the problem is due to lack of sufficient WDNR staffing to process permits.
  - Timeliness for getting permits appears to be completely out of line with other states.
  - Policy issues at the WDNR make it very difficult to get the job done, and usually result in high capital expenditures.
  - Need to push proactively for fast track permitting.
  - Streamline permitting to allow greater use of coal.
  - Difficulty dealing with the WDNR bureaucracy in Madison.
- Regulatory framework to support the implementation of emerging technologies. (i.e. permit for air emission relating to new combustion technologies or new fuel use)
- Need standardized requirements for reporting of Volatile Organic Compound (VOC) emissions on raw materials and like production units.
- Need innovative state and federal programs that will establish environmental and energy goals for the industry and eliminate regulatory barriers to achieving them.
- Allow permit credits for innovative technology applications (provide motivation to change).
- Need consistent and aligned rules or regulations governing the environmental aspects of the industry
- Determine cost/benefit on rulings for run-off, environmental regulations. (Major issues are storm-water run-off and air emissions.)
- Establish how so called "pollutants" fit into the natural system.
- State emissions regulations should match but not exceed federal regulations.
- Rules should apply through life of asset.
- Need to encourage innovative research to support infrastructure improvements.
- Need to develop and rewrite environmental policy from a command and control philosophy to a policy based on accomplishment.
- The WDNR and companies need to work together to better understand what the real problems are with permits and try to resolve them. It is important to standardize the process throughout the state.

- One way to speed up the process is to have automatic approval in x number of days if no action is taken by WDNR.
- Develop programs that give industry ownership and incentives to excel in environmental issues.
- Consider a hybrid version of International Standard Organization (ISO) 14000 environmental certification. Such a system, if mutually agreed upon, stands to take industry out of a defensive posture with regards to their environmental track record/history and gives them a chance to be proactive in policy development improvement and implementation.

Business
- There is a disproportionate negative impact of increased regulations on small mills. (High labor and capital costs)

Capital
- High capital costs to meet environmental concerns.

Technical
- Benchmark the permit processes used by other states.
- Need science-based regulations: facts and data to guide effluent quality requirements.
- Need to determine full range of permits involved, air, water, VOC, wetland, etc. and see if any of the permit processes used are examples of success.

Education
- Promote the triple bottom line for industry accounting.

PERCEIVED WISCONSIN BUSINESS CLIMATE:

Policy
- Wisconsin doesn't give forestry the kind of attention that it deserves.
- Feeling by some that Wisconsin is anti-business.
- The industry has advocacy groups but no real middlemen to help solve the complex issues facing forestry businesses.
- Wisconsin does not see itself as a manufacturing state.
- Other states are perceived to be pro business and work to make business feel welcomed.
- Current regulations hinder joint co-generation projects. (Viewed as public utility, regulations increase exponentially.)
- The state Family Medical Leave Act (FMLA) needs to be standardized to the Federal Act.

Business
- Need for sharing information on technology advancements—working together as an industry.
- Need more positive public relations and community support.
- It is believed that Wisconsin is less generous with economic polices than other states.
- Need aggressive programs to find means to implement new ideas.

Technical
- It is important that Wisconsin benchmark their present polices with other states such as Michigan, and Minnesota.
- Important for the Center for Technology Transfer (CTT) and the Forest Products Laboratory (FPL) to work together.

Education
- Need for a single organization in the state that could handle the problems and questions of the forest industry.
- Need to develop a comprehensive program to increase awareness of the importance of the wood industry to the state in terms of jobs, tax revenue, community stability, forest health, clean water, wildlife and recreation.

TAXATION:

Policy
- Government subsidies and tax breaks vary between states and countries.
- Need exemption from sales tax on fuel and electricity used in manufacturing.
- Need to implement a single sales factor for corporate income tax apportionment.
- Provide solutions to include investment and educational incentives.
- Provide incentives that encourage new research and development—& D).

- Provide incentives to phase out obsolete or inefficient capacity.
- Provide incentives to existing businesses—taxation, labor support/credits, investment/technology, tariffs / supports.

Business

- Benchmark taxes against other states.
- Benchmark taxes against other countries that compete.

Capital

- Industrial revenue bond investments to build new mills can negatively affect existing mills.

Education

- Complete a study and report on the use of the Wisconsin Forestry mil tax and its positive and negative impacts.

FOREST RESOURCES:

Policy

- Need to assure plentiful and suitable timber or other fiber resources for the state industry.
- Fragmentation of the forest is happening at an alarming rate. This affects ability of businesses to procure raw materials. Lack of available wood supply.
- Need to reduce concentration of excessive material in overstocked forests.
- Present legislation has pushed farmers to abandon programs that are aimed to help forestry. Farmers are once again allowing their cattle to graze in the timberlands. (Limited grazing may be helpful if properly done. Bacterial infections are one problem associated with grazing)
- Long time (up to 12 months) taxpayers have to wait for certain tax credits or payments.

Business

- Complete a study of the increase in Wisconsin growing stock, the limits of its availability for utilization, and options.

Technical

- Provide help to Non-Industrial Private Forest (NIPF) landowners in the development of forest management plans and assistance in working with loggers and lumber companies to meet their forest plans.
- Non-industrial woodland is not properly managed.

Education

- Many private woodlands are not managed because of lack of trained foresters to make or approve forest plans. (Continuing need to educate landowners)
- Need to bridge consumer/public disconnect with science/study findings on forest management.
- Review the literature and provide a report on possible utilization of each species. For example the use of saw-dry-rip to utilize species that are hard to dry without severe defect. This would help industry to better utilize species that are not commonly used, but are in abundance.

ENERGY:

Policy

- Need for reliable energy supplies over time.
- Need to improve the electrical grid inside state and linking Wisconsin to other states, especially to the west. Transmission capacity is becoming a critical concern with deregulation.
- Need regulated pricing mechanism for fixed and interruptible power.
- Price volatility: Improve control of natural gas pricing.
- Need ability to access open energy markets.
- Need to open the generation market to Independent Power Producers.
- Need reliable sources of fuel—renewable, less fossil fuel dependent.
- Dam removal issues: balance environmental improvements versus renewable energy. Hydropower needs to be revived.
- Use of artificial wetlands for final finishing of wastewater treatment (low energy and tertiary treatment).

Business

- Need to replace oil and gas for steam production by wood, wind, solar, or other renewable energy source, for building and process heating.

- Ability to remain energy competitive while utilizing aged steam-generating systems.
- Need to review the WDNR report on good sites for district heating and follow-up on opportunities.

Technical

- Need clean, economical energy source to produce steam and electricity.
- Need to develop flexibility in fuel uses to optimize facility costs and environmental factors.
- Need to develop a portable cogeneration unit for utilization of dead and down material in fire prone forest areas. Work is progressing on a 1 MW unit that is portable. It should be ready in three years. It will take 3 semis to transport the unit.

———

Mr. WALDEN. Let us go now to Mr. Coston. Sir, thank you for—and perhaps, Mr. Rehberg, did you want to make any opening comments?

Mr. REHBERG. I would just like to welcome Tom. It is nice to see you again. He has been dogged in his work with the RC&D down in Bitter Root. As you know, the fires of 2000 were not particularly kind to the Bitter Root area of Montana. Thank you for sticking with this project as long as you have and I think you will be impressed with what they have been able to accomplish.

STATEMENT OF TOM COSTON, FUELS FOR SCHOOLS COORDINATOR, BITTER ROOT RESOURCE CONSERVATION AND DEVELOPMENT AREA, INC., HAMILTON, MONTANA

Mr. COSTON. Thank you, Denny. We had the pleasure of a visit from Congressman Rehberg very early on in this and we have enjoyed his support ever since, he and his staff.

My name is Tom Coston. I am the Coordinator for the Fuels for Schools program. I work for the Bitter Root RC&D. This has been a team effort, also, and our partners in this have been the Forest Service's Northern and Intermountain Region, it has been the Forest Products Lab, it has been the State Foresters of the five-State area that the Northern and Intermountain Regions cover, and it has also been the Biomass Energy Resource Center, which is located in Montpelier, Vermont.

I kind of wondered when I first got in here if I wasn't in the wrong place, because everybody was talking about how to use the biomass to build something. All we want to do is burn it up, and that is what we are about, really. It is about using waste wood, wood chips, to fire boilers to heat public buildings. The emphasis is on schools and the emphasis is on wood coming from fire hazard reduction operations, and you have already talked about what those operations are.

As Congressman Rehberg said, we got into this as a result of the summer of 2000. What that did, it woke up the people in our valley, at least, in our area, that something needed to be done to lessen the potential of a repeat of that type of year. Hazard reduction logging or thinning became the thing that we tried to do.

That created a little problem on its own and that was what to do with all this unmerchantable material from the standpoint of forest products. We kind of fell back on what has been going on in New England and the Lake States, Eastern Canada over the last 20 years. There has been quite a movement toward using wood chips for heat. It is a plentiful, inexpensive, renewable, non-fossil energy source that had the promise of being able, for the facilities using it, to be able to cut their heating costs significantly.

So the partnership was able to set up a pilot project in the town of Darby, Montana, about 15 miles up the road from where I live, and it was made possible by a grant using a Forest Service State and Private Forestry Economic Action program fund, which you folks and your colleagues over in the Senate made possible. Our pilot project went on line last November, or last October 30, and has run successfully and did a good job of demonstrating a forest biomass heating system throughout the year. I suppose the bottom line was that it saved Darby public schools over 50 percent in their heating cost.

Our present situation is that in Montana, and I should say that Montana has a leg up on this because we got a couple of years head start over the other four States, but in Montana, we have two additional projects that are under construction right now and will be online this fall. Idaho has two projects that are well along in planning and design and they hope to be able to go to construction before the year is out, also. Nevada has one and North Dakota has one. Those, plus Utah being the other States that are in the Northern, if you aren't familiar with what is in the Northern Region and the Intermountain Region.

All the States have a number of projects that are waiting in the wings. The biggest obstacle we have right now is the high up-front cost of a biomass heating system. The top line of equipment costs around $600,000. Our objective, our long-term goal in this as far as Fuels for Schools goes, is to get out of the grant business and let the economic benefits of biomass heating carry the program itself. Before we can do that, though, we have got to get the cost, the capital investment cost down, and that right now is the main plank in our platform.

There are smaller, less expensive equipment out there. It simply hasn't been tried and tested in our area for our particular use. We have a contract right now with the Biomass Energy Resource Center where they are on a fast track to identify the best quality of these systems and we hope to have at least one in operation before the end of this year.

I think I probably used up the majority of 5 minutes. I would like to just say that we appreciate being offered the opportunity to appear here. We are kind of meat and potatoes compared with most of what you are hearing, but it is a pleasure and we appreciate your interest.

I think you are to be commended on the work you have done on getting the Healthy Forests Restoration Act into law and we hope that you are able to implement it completely. We also hope you can continue to support the Forest Service's Economic Action Program funds, because that is what made our program possible.

I would be glad to answer any questions.

Mr. WALDEN. Thank you, Mr. Coston. We appreciate your comments today, certainly, and they have been most intriguing about practical applications here.

[The prepared statement of Mr. Coston follows:]

**Statement of Tom Coston, Fuels for Schools Coordinator,
Bitter Root Resource Conservation & Development Area, Inc. (RC&D)**

FUELS FOR SCHOOLS PROGRAM

My name is Tom Coston. I'm from Hamilton, Montana and I represent the Bitter Root Resource Conservation and Development Area, Inc. (RC&D). I want to talk about "Fuels for Schools", the name coined for a program which advocates using forest biomass as fuel to fire boilers to heat schools and other public buildings.

As all of you know, the past three summers have been severe wildfire seasons, particularly in the West. In 2000, Montana was perhaps the hardest hit and the Bitterroot Valley was the epicenter of that fire activity. Over half a million acres burned, along with many homes and other structures. Many others had fire at their doorstep and were saved by tremendous effort by fire crews and at great expense to the taxpayer.

This did, however, create an awakening of the need to do something to reduce the threat of future fires, such as removing enough of the fire hazardous material to make fires more manageable. We are talking about all land ownership, particularly those along the wildland-urban interface, a term we never heard before this. Many people and many land ownerships moved in the direction of fire hazard reduction.

The immediate problem was what to do with the large volume of logged or otherwise removed material—forest biomass became the term—most of it unmerchantable from a forest products point of view.

In our area the U.S. Forest Service (Bitterroot National Forest) and the Montana State Forester assumed leadership of an effort to find ways of utilizing this mostly small-diameter material. They enlisted the aid of my organization, the Bitter Root RC&D, a non-government, non-profit organization, whose charter is to assist our communities and elected officials in affecting conscientious natural and human resource decisions.

While effort was made in numerous directions to utilize this material, the one we are talking about today is the use of chipped waste wood as a fuel. We found that in New England, Eastern Canada and the Lake States over the past 20 years there has been an expanding interest in using wood chips to fire boilers to heat buildings—mostly schools. The technology for completely automated systems had been perfected and the bottom line was that heating costs could be significantly reduced—50 percent not being unusual.

Waste, or cull wood, is run through a chipper to produce a usable fuel, then fed by automated systems to a burner and boiler to heat water or make steam. Beyond that point the heating systems are the same as other conventional ones commonly used, such as fuel oil or gas.

The Forest Service Forest Products Laboratory (FPL) of Madison, Wisconsin and the Biomass Energy Resource Center (BERC) of Montpelier, Vermont helped set up a local pilot demonstration project. The Forest Service had funds available for grant assistance under the National Fire Plan. This "partnership" surveyed the local school districts and concluded that Darby was the best site available to demonstrate the operation and benefits of a biomass system for several reasons, such as good community support for the trial and the fact that Darby had the greatest potential to demonstrate savings. The fires of 2000 burned all around Darby, and fuel oil to heat their 3-building campus was costing about $60,000 per year.

You and your colleagues in the Senate made funds available through the National Fire Plan using the Forest Service Economic Action Program. A grant (actually two grants over two years) was assured to fund the $870,000 construction. The agreement called for Darby to monitor and evaluate the operation, including all costs, for a two-year period, and to make the operation of the biomass heating system available for demonstration to the interested public.

The system was fired up last October 30, 2003 and ran successfully throughout the school year. The previous year fuel oil to heat the Darby complex, as said before, cost about $60,000. Actual cost of 640 tons of wood chips for the school year just concluded was $18,500, and about $11,000 of fuel oil was burned in September and October and as backup, bringing the cost to about $29,500. The school year ended two weeks ago and the costs are still being evaluated, but it appears reasonable to expect a full school year of wood chips will cost about $20,000. The school was able to utilize the savings for other priorities in their educational charter

The State and Private Forestry program of the Forest Service has expanded the Fuels for Schools program to cover the 5-state Northern and Intermountain Region area—Montana, Idaho, Nevada, Utah and North Dakota. The State Foresters manage the program in their respective states. Interest is very high throughout the area and progress is being made.

Assistance grants are available and are now being structured as generally covering 50% of the overall costs of converting to a biomass system with the school (or other facility) financing the remaining 50%. In Montana, low interest state-sponsored "intercap" loans are so far the preferred vehicle. Feasibility studies are done for each candidate with a key ingredient being the ability of the school to pay back its loan over a 10-year period with fuel cost savings. The idea is to make the conversion cash-positive the first year.

In Montana, two other demonstration sites (Victor and Phillipsburg) are now under construction and will be operating this fall. A fourth, Eureka, is securing its funding and hopes also to begin construction.

In Idaho, two communities are committed to going forward with the demonstrations and are well into planning. Ely, Nevada has made a decision to proceed and is also in planning and design. In Bottineau, North Dakota, Minot State University is committed to a demonstration project and is going ahead. All of the states have a number of other sites that are "waiting in the wings."

The long-range goal of Fuels for Schools is to adequately demonstrate the benefits of biomass systems with the help of assistance grants, and after a reasonable time, to allow the economic benefits of conversion to provide its own momentum, with the institution and the private sector providing financing.

The major obstacle right now is the high up-front cost of a fully-automated biomass system—about $600,000. A rule of thumb has evolved that a school, or other facility, must be heating between 50,000 and 100,000 square feet, and incurring a proportionately large heating bill in order to generate enough savings to make conversion pay out over a 10-year period. Most schools and other facilities are under that size.

Fuels for Schools is currently making a major effort to reduce the capital investment cost. Smaller "semi-automated" systems are available. They are as yet untested in our area. The major differences are smaller boilers, a much smaller boiler building, a "hopper" fuel storage design that must be mechanically filled periodically, and a cost of less than half that of the larger systems.

We plan to identify the best quality of such systems and to install at least one as a demonstration yet this year.

I think we collectively have a good program and, as with most worthwhile things, much work remains to be done. If all four of the Montana sites were presently active, we would only be using some 2,500 tons of material annually—the annual thinnings from about 200 acres. Someone estimated that in our 5-state area hazard reduction treatment annually results in well over 2 million tons of material.

My organization appreciates your support and lauds your effort in passing the Healthy Forests Restoration Act. This hearing demonstrates your commitment to follow through. We would like to see the Act fully implemented, as well as continued support for the Economic Action Program funds which have enabled our program to go forward.

We are grateful for your invitation to appear here and if there is any way we can assist in furthering this effort, would appreciate being called upon.

If you have questions, I would be pleased to respond to them.

———

[An attachment to Mr. Coston's statement follows:]

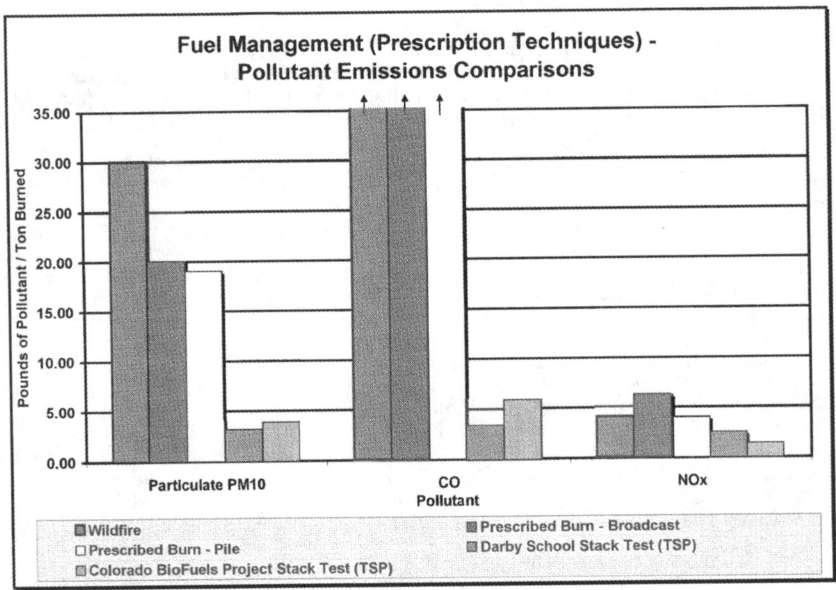

Mr. WALDEN. Let us go now to Lynn Jungwirth. Welcome. Thank you. We look forward to your comments, as well.

STATEMENT OF LYNN JUNGWIRTH, EXECUTIVE DIRECTOR, THE WATERSHED RESEARCH AND TRAINING CENTER, HAYFORK, CALIFORNIA

Ms. JUNGWIRTH. Thank you for having me here today. I think I represent public land communities. Most of us are distressed economically and most of us are surrounded by forests that are unhealthy.

So in my community about 10 years ago, we raised the issue of fire and began working collaboratively with anybody who would work with us to try to figure out how to solve that problem. One of the things we first recognized was that it cost too much to bring that stuff out of the woods and there are no local markets for it. The definition of biomass appears to be whatever is not merchantable in your local area. In San Bernardino, it is a 30-inch log. In parts of Oregon, it is five inches and under.

In our area, we had a sawmill that could still manage to manufacture things that were nine inches in diameter and above, but the smaller stuff needed to come out. So because of the Economic Action Program, we were able to be a local partner to the Forest Products Lab. They also got their funding for the Technology Marketing Unit through the Economic Action Program. And we began working on this project.

So the first experiments we did removed this biomass at a cost of around $20 a ton. So that was a sunk cost. That was a lost cost. We hauled that down to a sort yard, sorted it for highest and best use, shipped some of the stuff off to the local mills, some of the bigger stuff, but the smaller stuff we began monkeying around with.

The first thing we did was turn it into a commodity, a two-by-four and a two-by-six. We could sell it for $200 a thousand. It still didn't pay its way out of the woods. We had it graded. We worked with the Forest Products Lab. We discovered that that suppressed material was of very high value, and so as structural lumber, the value increased up to $350 a thousand. But we were still making two-by-fours and two-by-sixes.

Some of the local entrepreneurs looked at that and said, my goodness, this is remarkable suppressed fir. It looks like old growth. Let us use it for flooring and paneling. You kiln dry it, you mold it, and all of a sudden it is on the market for $1,250 a thousand. Now it has a value at the stump.

As biomass for a biomass plant in our community, with the current pricing structure, that material is worth $7 a ton if we had a biomass plant in my community, which we no longer have because our mill closed down and their co-gen plant left But we have done the numbers and worked with industry. We could support a 13-megawatt plant, a 13-megawatt plant that could burn the residual off of the small log processors that would be co-located around that plant with the markets that we have developed, both for flooring and paneling, post and poles, tepee poles.

We sold 25-foot-long, I can't call them logs, I don't know what they are, trees, an inch in diameter on the top, two-and-a-half inches on the bottom. We found markets for those. There are markets for those.

The sort yard with the colocated processing as a stand-alone doesn't quite work, and the reason it doesn't quite work is we have that residual to get rid of, the mill waste. The biomass plant as a stand-alone can only pay $7 a ton. It cannot pay for that stuff coming out of the woods. But the small log plant can pay $40 a ton and it could pay for the stuff coming out of the woods.

When you take the small log plant and you put that together with a biomass plant and you put that together with a kiln and somebody who could buy the downstream heat from the biomass plant, now you have a system that works economically in a rural community. We don't have natural gas available. Most of your mountain communities don't. People who need to manufacture things that need a cheap source of heat, electricity isn't it. They could pay a little bit back to the biomass plant. It makes the numbers work. That is the proposal we are moving forward with.

The issue of national forest management and rural community vitality has to stay uppermost in your mind as you work on forest health. You need us in the woods. It will reduce your cost of suppression. You will have crews who know the landscape, who can respond to fire. You will have skilled people who can deal with this material. Everywhere across the West, little communities like mine are finding these value-added solutions and really creating high-value products. They are not commodity products.

I encourage you to encourage us to keep doing that experimentation and support the programs that have supported us. Thank you.

Mr. WALDEN. Thank you. Your comments are most helpful.

[The prepared statement of Ms. Jungwirth follows:]

48

**Statement of Lynn Jungwirth, Executive Director,
The Watershed Research and Training Center, Hayfork, California**

The Watershed Research and Training Center is located in a very small public
land community in the heart of the Shasta-Trinity National Forest. Since 1996, we
have been working with local businesses, the Economic Action Program of the
Forest Service, and the Technology Marketing Unit of the National Forest Products
Lab to develop infrastructure for the removal and use of hazardous fuels.

This hearing is dedicated to the potential in "bio-mass". In our experience, the
definition of biomass depends upon the utilization capacity in an area. In my county, any soft wood tree under 9" in diameter is considered sub-merchantable and
therefore good only for fuel for biomass fired electrical generation. Our goal was to
find the highest and best use for these smaller softwoods and the underutilized
hardwoods in our area. We saw this value-added approach as the best way to create
jobs in our economically distressed community while taking care of forest.

To that end we created a worker re-training program, developed specialized equipment for fuels removal and wood processing, and opened a business incubator for
value-added wood product entrepreneurs. We have also collaborated with others to
create strategic community-based fire plans in the 17 communities in our 2.1 million
acre rural county.

The Watershed Center's programs have reduced fuels on over 1500 acres of public
and private lands using hand crews, ground-based equipment on flatter ground, and
skyline yarding on steep ground. We believe that skilled crews and more efficient
equipment are important pieces to the forest health puzzle.

The Watershed Center has been the hub of needed research and development for
small diameter timber and under-utilized west coast hardwoods; our efforts have
succeeded in helping local businesses manufacture and market tee-pee poles (a pole
25 ft. long, 1.5 inches on the top and 2.5 inches on the butt), fence poles, roundwood
for furniture and fixtures, flooring, paneling, and store fixtures. By taking this integrated approach we have created over 25 jobs in value-added businesses and run
a 9-person fuels crew.

Along the way we have worked with industry consultants on various feasibility
studies for large-scale wood processing operations, sort yards, and bio-mass fired
electrical generation plants. Some of those lessons are incorporated in my testimony
today.

The Watershed Center also works will many public land communities throughout
the west, helping them create their fire plans and their efforts to capture social and
economic benefit from fuels reduction and forest restoration projects. With Sustainable Northwest and Wallowa Resources in Oregon, we have fostered a marketing
association for small businesses making value-added products from the by-products
of forest restoration and fuels reduction. This association is over five years old and
is called "The Healthy Forests, Healthy Communities Partnership".

I would like to take this opportunity to share some of the lessons we have learned
from our research and development programs in small diameter and hardwood utilization.

I will also provide a few comments on federal programs that have been critical
in the development of products, processes, and businesses that utilize hazardous
fuels for value-added products.

High Value Products

Key Points in Considering Biomass Utilization
- Biomass utilization must facilitate and complement restoration activities, not
  override restoration needs with high input demands. The scale of biomass
  plants needs to be consistent with ecosystem capacity and tailored to restoration
  objectives. Maintaining social support is critical.
- Federal programs have been proposed to develop and establish biomass utilization centers and subsidize transportation costs. This money should be used to
  develop diversified forest products sectors (including uses beyond energy generation) at the community level and not to subsidize large centralized plants
  with little stake in forest-dependent communities.
- All economic opportunities for biomass utilization should be targets of government support, not just biomass energy generation. Using biomass for power
  should complement and diversify the approaches to small diameter wood utilization. Stand-alone biomass energy generation, while allowing for the utilization
  of a large volume of material, entails the creation of the fewest jobs of all biomass utilization approaches. Co-locating value-added processors with a wood
  fired electrical plant improved the economics of all the plants.

49

- Biomass transportation subsidies will help to offset the costs involved, but may also act to increase the reach of large facilities to the exclusion of small businesses. Encouraging small and micro facilities will require more focused subsidies to create the greatest benefits for rural communities, and to encourage entrepreneurialism, research and innovation. Short-term subsidies should help foster the development of long-term self-sustaining uses and new technologies.
- Local context is essential to appropriately choosing and siting biomass utilization facilities: what are the restoration needs (and biomass supply) that can drive facility development; what combination of technologies will add the most value to biomass and create the most jobs in the area; what experiments can be supported locally to advance regional knowledge of opportunities for innovative utilization?

An example

In 1996, when we started our small diameter utilization program, fuel for the biomass plant had a market value of $11/ton. A truckload weighs roughly 25 tons. Market value was then $275/truckload delivered to the biomass plant which was located many miles away in the valley. The cost of the haul was $330 dollars. Our recent study shows a value of only $7 for fuel if a stand-alone biomass plant was operating in our town. The difference is the 7 cents per kilowatt hour the valley plant is paid under an old contract and the 5.5 cents per kilowatt hour the new plant would operate under. At those market values the cost of extraction and chipping and most of the transportation costs would not be covered. Costs to the landowner, the Federal Government in this case, would be about $20-$30/ton, depending on the terrain and the haul distance.

We thinned a stand of suppressed Douglas Fir. It averaged 7.5 inches in diameter and was about 110 years old. We brought the material to town, sorted it for best use and began to make things out of it. The first things we made were construction lumber, 2x4s and 2x6s. The value rose to $200/mbf after processing or $20/ton. We then graded the lumber, found it was of excellent quality, and the value rose to $350/mbf. Since it was beautiful wood, we turned it into flooring and cabinet framing and the value rose to $1250/mbf. Every dollar increase was tied to a job. Today we can safely pay $45/ton for sub-merchantable hazardous fuels. Most work pays for itself.

From a per acre cost of $700 to $1200 dollars to the taxpayer, we moved to a break even on most acres and an average of $300 on the steepest, most expensive ground. We still have waste product from the processing that needs to find a market. If we had a small biomass generating facility in our town, co-located with our small wood processing facility, we would greatly improve our economic picture and even more costs of treatment could be covered.

Our solutions were all low tech, non-traditional, small-scale. If the assistance we received had not been comprehensive (FPL, demos, marketing) and had simply been a $20/ton subsidy for hauling biomass fuel, these innovative, higher value products and markets would have never been developed.

Federal Programs critical to bio-mass utilization and rural community development:

- National Fire Plan Economic Action Program under State and Private Forestry in the U.S. Forest Service Budget. This program has been zeroed out of both the President's and the Congressional budget for the past two years and is zero in the proposed 2005 budget. It has supported most of the successful bio-mass value-added projects I know about in the west.
- Economic Action Program base program under State and Private Forestry in the U.S. Forest Service Budget. This program has supported many utilization development projects around the country and is seriously underfunded.
- The Technology Marketing Unit of the Forest Products Laboratory in Madison. They offer the best and the most accessible technical assistance to businesses and communities in this country. They struggle for funding every year and deserve your support.
- The Stewardship Contracting mechanism now available to U.S. Forest Service and BLM will allow 10 year contracts, alternative funding mechanisms (goods for services/retained receipts), lower costs (designation by prescription/description) and social processes to insure continued support (collaboration and multi-party monitoring). This approach will take time to perfect but is the most promising policy tool for fuels reduction.

Utilization of hazardous fuels was and is a key to the National Fire Plan and other forest health efforts. Today, with the downsizing of the Economic Action Program (EAP) of the Forest Service and the elimination of the National Fire Plan EAP and the loss of community assistance dollars (Community and Private Lands Fire

Assistance) the Forest Service is clearly walking away from the utilization commitment. BLM has no dedicated program for utilization. I urge the subcommittee to end this disconnect. A small investment in utilization will reap huge benefit to the taxpayer and rural communities.

The high value products from hazardous fuels are not limited to wood products, but also include an unprecedented social agreement to manage public lands. This social consensus happened because of on-going collaboration at the local level. I urge this subcommittee to keep the collaboration envisioned in the National Fire Plan alive. Working together for the good of the forests and the good of the public land communities is the best strategy to insure forest health.

SUSTAINABLE NORTHWEST
620 SW MAIN, SUITE 112, PORTLAND, OREGON 97205
503.221.6911 FAX 503.221.4495 WWW.SUSTAINABLENORTHWEST.ORG

COMMUNITY-BASED PERSPECTIVES ON BIOMASS
BRIEFING PAPER

Forest Restoration and the Problem of Biomass

Biomass refers to living, or recently living, woody material that cannot be economically processed through traditional means. Western forest restoration treatments often require the removal of large numbers of trees that are either too small, too decayed, or too misshapen to be used as sawlogs. Biomass removal is essential to the restoration of Western forests: small diameter wood accumulation is a major contributing factor to catastrophic wildfires, and thickets of small diameter trees often contribute to a general lack of forest health and resilience (including low growth rates, insufficient soil humidity, and outbreaks of disease).

Biomass can be marketed to create fiber products such as paper and cardboard, but weak prices or fire effects on the wood can eliminate this possibility. Even where there is pulpwood demand, harvest and transport costs (in addition to facility operation costs) tend to be too high to allow economic returns in an unsubsidized market. Economic returns for biomass are significantly affected by distance: with demand low to begin with, long hauling distances significantly diminish potential profits.

Most often, biomass is chipped, pile-burned on site, or buried in landfills, generating significant costs and providing nearly no social or ecological benefits. Due in part to the high cost of dealing with biomass, important restoration needs across the West have consistently gone unmet: ironically, the material that is often the most important to remove as part of restoration treatments is also the least commercially viable. Finding economic uses for biomass can significantly support the implementation of forest restoration activities, while providing much-needed economic benefits rural communities.

Biomass Utilization

Biomass utilization entails putting this material to some kind of commercial use. The term often is associated with energy generation facilities ("biomass energy generation"), but it refers to a whole host of uses for small diameter wood, such as roundwood building materials, posts and poles, forest products such as flooring and paneling, and other innovative uses such as erosion control structures. Adding value to small diameter material through processing and manufacturing—whether it be into forest products or energy—may create sufficient economic returns to overcome the costs of biomass removal in forest restoration activities.

In the past few years, ongoing population growth and increasing demand for electricity in Western states, along with recent swings in Western energy prices (e.g. California's electricity price fluctuations in 2001), has generated great interest among forest communities in linking forest restoration activities to biomass electricity generation. Biomass can be converted to consumable energy through several types of facilities. Qualifying Facilities ("QF's") convert biomass to electricity through a steam process similar to coal-based electrical generation (biomass used on its own or with other combustibles to fire a steam plant that in turn drives electrical generators). Co-generation facilities ("co-gens") produce electricity in addition to other outputs, the most common being heat or steam used in lumber kilns. Other classes of biomass facilities are capable of converting the material into fossil fuel substitutes such as ethanol or other transportable fuels. Biomass energy is an alternative to non-renewable energy sources, such as fossil and nuclear fuels, and is generally considered "green" energy (though there is some debate given that it can generate polluting by-products).

Recent legislation has authorized funds to help subsidize transportation and utilization of biomass, and to fund continuing research and development of biomass technologies:

- Biomass utilization is specifically encouraged by the National Fire Plan: "Because much of the hazardous fuels in forests are excessive levels of forest-based biomass—dead, diseased and down trees—and small diameter trees, there are several benefits of finding economical uses for this material, including helping offset forest restoration cost; providing economic opportunities for rural, forest-dependent communities; reducing the risks from catastrophic wildfires; protecting watersheds; helping restore forest resiliency, and protecting the environment." (p. 25)
- The Biomass Research and Development Act of 2000 (Title 3 of the Agricultural Risk Protection Act of 2000, P.L. 106-224) allows entities (including nonprofits) to compete for federal grants and contracts associated with biomass research.
- Section 9006 of the 2002 Farm Bill (P.L. 107-171) authorizes federal grants and loans to farmers, ranchers, and rural small businesses to purchase renewable energy systems, and section 9010 authorizes payments to producers of bioenergy (biodiesel or ethanol). The FY2004 Farm Bill (P.L. 108-199) appropriated $23 million to fund these provisions.
- Section 201 of the Healthy Forests Restoration Act of 2003 (P.L. 108-148) expands the scope of these grants to include research on thinning, harvesting, transportation, pricing, and curricula development. Section 203 of HFRA authorizes grants to owners and operators of biomass facilities, including wood-based product facilities, and authorizes funds to this end.

Scale and Adding Value

Biomass energy generation facilities can range from very small, generating enough power or heat for use in a single building (such as a school or mill) to generating enough electricity to power tens of thousands of homes. "Micro" facilities are those generating less than one megawatt of power; "small" facilities are those producing 1-10 MW. Establishing a biomass facility requires a dependable, sustainable supply of biomass within the nearby area (25 to 75 miles). When looking at the combined needs for forest restoration and rural economic development in general, maximum social and environmental benefits will likely result from many smaller units distributed among forest-based communities, rather than fewer, larger facilities.

Supporting and siting biomass utilization facilities must be done with consideration of many local factors. Even at smaller scales, building biomass generation facilities can raise concerns about developing unsustainable demands for biomass materials, creating pressure to "deliver material" rather than to restore forests. Biomass generation facilities must also be weighed carefully against other potential uses of biomass that can either complement, or surpass generation facilities in their ability to provide rural employment opportunities through value-added processing. Siting and planning of biomass utilization facilities must be closely coordinated with local forest restoration goals and a community's particular economic circumstances. Given that the focus on biomass utilization and forest restoration is a recent one, every effort should be made to promote diversity and experimentation as a short-term path to identifying successful long-term utilization solutions.

Key Points in Considering Biomass Utilization

- Biomass utilization must facilitate and complement restoration activities, not override restoration needs with high input demands. The scale of biomass plants needs to be consistent with ecosystem capacity and tailored to restoration objectives.
- Federal funds are coming online to develop and establish biomass utilization centers and subsidize transportation costs. This money should be used to develop diversified forest products sectors (including uses beyond energy generation) at the community level and not to subsidize large centralized plants with little stake in forest-dependent communities.
- All economic opportunities for biomass utilization should be targets of government support, not just biomass energy generation. Using biomass for power should complement and diversify the approaches to small diameter wood utilization. Stand-alone biomass energy generation, while allowing for the utilization of a large volume of material, entails the creation of the fewest jobs of all biomass utilization approaches.
- Biomass transportation subsidies will help to offset the most prohibitive costs involved, but may also act to increase the reach of large facilities. Encouraging small and micro facilities will require more focused subsidies to create the greatest benefits for rural communities, and to encourage entrepreneurialism,

research and innovation. Short-term subsidies should help foster the development of long-term self-sustaining uses and technologies.

- Local context is essential to appropriately choosing and siting biomass utilization facilities: what are the restoration needs (and biomass supply) that can drive facility development; what combination of technologies will add the most value to biomass and create the most jobs in the area; what experiments can be supported locally to advance regional knowledge of opportunities for innovative utilization?

For more information contact:

Jesse Abrams, Wallowa Resources,541-737-3888;
  Jesse.Abrams@oregonstate.edu
Jim Walls, Lake County Resources Initiative, 541-947-5461;
  jwalls@gooselake.com
Jim Jungwirth, Jefferson State Forest Products, 530-628-4206;
  jim@jeffersonstateproducts.com
Nils Christoffersen, Wallowa Resources, 541-426-8053;
  nils@wallowaresources.org
James Honey, Sustainable Northwest, 503-221-6911 x 106;
  jhoney@sustainablenorthwest.org

———

Mr. WALDEN. Let us go now to Jason Drew. Thank you and welcome. We are delighted to have you with us today.

STATEMENT OF JASON DREW, DISTRICT MANAGER, NEVADA TAHOE CONSERVATION DISTRICT, STATELINE, NEVADA, ON BEHALF OF THE NATIONAL ASSOCIATION OF CONSERVATION DISTRICTS

Mr. DREW. Thank you, Mr. Chairman, committee members. On behalf of America's conservation districts and the National Association of Conservation Districts, I am pleased to provide you with our insight on the role conservation districts play and can play throughout the country on hazardous fuels reduction, woody biomass utilization, and regulation planning.

Hazardous fuels buildup is a serious threat nationwide. Conservation districts strongly support efforts to reduce hazardous fuels buildup, develop new and innovative technologies to use woody biomass, and to educate the public about proper forest management.

The decline of the forest industry in the West, as we have heard from some of our other panelists, contributes to the problem by removing many business options for utilizing woody biomass. Distances from markets and the high cost of transportation make utilizing woody biomass even more difficult.

Conservation districts applaud the Congress for its quick action on the Healthy Forests Restoration Act. Its funding and implementation through the National Fire Plan provide opportunities for local communities and organizations, including conservation districts, to become engaged in fuels reduction projects and education. Commitments from Congress and the Administration to this end is crucial to the success of this effort.

Conservation districts and resource conservation and development councils, as we heard earlier, already have in place a number of cooperative agreements with Federal land management agencies to promote and improve the utilization of woody biomass in order to reduce catastrophic wildland fires and restore forest, woodland, and rangeland health.

In my conservation district, the Nevada Tahoe Conservation District in Stateline, Nevada, forest conditions in areas surrounding Lake Tahoe are indicative of many areas in the Western U.S. experiencing an accumulation of excess fuels, leading to reduced resistance to wildfire, disease, and insect infestations. These large quantities of biomass are not merchantable as wood, often, and through other manufacturing industries. However, utilization of this biomass for energy offers a potential economic use for this material which would help reduce fuel loads.

We recently completed a woody biomass resource and technology assessment for the Lake Tahoe Basin. The study quantifies the Basin's biomass resources and costs, analyzes biomass energy technology performance characteristics, assesses local opportunities for using the material, and summarizes the results of initial planning on a pilot project conducted in conjunction with the Lake Tahoe Unified School District. The study showed that there are opportunities for small-scale biomass energy systems to be deployed in the Lake Tahoe Basin, which is some of the most heavily regulated land in the world.

As a result of the biomass feasibility assessment sponsored by my district, the Lake Tahoe Unified School District is pursuing further funding to purchase a co-generation boiler system to be deployed in the local high school, and this is a project separate from the one mentioned earlier. Biomass to run the new system will be supplied by Basin land management agencies from fuels management projects. I have attached a copy of the executive summary of this assessment to my written statement.

In your Congressional district, Mr. Chairman, the Deschutes Soil and Water Conservation District received an $89,000 National Fire Plan Community Assistance Grant in 2001 to implement an innovative project that turns woody biomass into compost. The grant enabled the district to implement the composting project for Sun River Utilities, which serves Sun River Lodge and Resort and about 4,000 homes in the Sun River development near Deschutes National Forest.

A landowner group was concerned about wildfire and undertook fuels reduction efforts in the lodge pole and Ponderosa pine forests which produced woody biomass from the treatments. The organic compost after utilizing that excess woody biomass was then sold or spread on Sun River golf courses. The organic compost and building supply composition in the area with volcanic soils that lack organic matter make it a valuable soil additive.

The district continues to focus on initiatives that turn woody biomass liabilities into assets. The district says it needs incentives and marketing capacity to demonstrate the value of that material.

Conservation districts believe efforts such as those I just described and other innovative projects offer tremendous opportunities to reduce catastrophic wildland fires and restore forest, woodland, and rangeland health. In fact, NACD recently entered into a cooperative agreement with the Bureau of Land Management, the Forest Service, and others to develop, promote, and improve woody biomass utilization. Other partners in this effort include the Interior Department's Bureau of Indian Affairs, the National Park Service, and the Fish and Wildlife Service.

Under this agreement, NACD is providing resource materials and information to local conservation districts to educate landowners and others on the issue. The goal of this initiative is to help increase public understanding of the social, economic, environmental, and aesthetic benefits gained by using woody biomass as a means of reducing fuel buildup on public lands. We believe more cooperative efforts such as this are needed. Involving local communities and landowners is an ideal way to ensure success of the Healthy Forest initiatives and the National Fire Plan and other wildland efforts.

We appreciate the opportunity to provide the Subcommittee with our views and I would be happy to answer any questions.

Mr. WALDEN. Thank you, Mr. Drew, and thank you, too, for your comments about the work being done by the conservation district that serves Sun River and the golf course there. I intend to do a personal tour of that on Saturday—

[Laughter.]

Mr. WALDEN.—to make sure that the riparian areas are properly treated and groomed.

[Laughter.]

[The prepared statement of Mr. Drew follows:]

Statement of Jason Drew, District Manager, Nevada Tahoe Conservation District, on behalf of the National Association of Conservation Districts

The National Association of Conservation Districts is the nonprofit, nongovernment organization representing the nation's 3,000 conservation districts, their 16,000 board members and 7,000 employees. Established under state law, conservation districts are local units of state government charged with carrying out programs for the protection and management of natural resources at the local level. Conservation districts work with a number of federal, state and other local agencies, as well as the private sector to provide technical and other assistance to millions of landowners and other partners to achieve this end. They provide the critical linkage for delivering conservation programs on nearly 70 percent of the private land in the contiguous United States.

In carrying their mission, districts work closely with the USDA's Forest Service, Natural Resources Conservation Service and the Interior Department's Bureau of Land Management to provide the technical and other help private landowners need to plan and apply complex conservation treatments on forest, range and other working lands.

On behalf of America's conservation districts, I am pleased to provide you with our insight on the role conservation districts play, and can play, throughout the country in hazardous fuels reduction, woody biomass utilization and forest planning.

Hazardous fuels build up is a serious threat nationwide. It threatens the viability of national forests, private forestlands—industrial and non-industrial and property in the wildland-urban interface. Excess woody biomass is exacerbated by the long-term drought plaguing much of the country and insect infestations, which in turn raises the danger of devastating wildfires that destroy wildlife habitat, communities and human life. Conservation districts strongly support efforts to reduce hazardous fuels build up, develop new and innovative technologies to use woody biomass and to educate the public about proper forest management.

The decline of the forest industry in the West contributes to the problem by removing many business options for utilizing woody biomass. Distances from markets and the high costs of transportation make utilizing woody biomass even more difficult.

Conservation districts applaud the Congress for its quick action on the Healthy Forests Restoration Act. Its funding and implementation through the National Fire Plan provide opportunities for local communities and organizations, including conservation districts, to become engaged in fuels reduction projects and education. Commitment from Congress and the administration to this end is crucial to the success of this effort.

Conservation districts and resource conservation and development councils (RC&Ds) already have in place a number of cooperative agreements with federal land management agencies to promote, and improve the utilization of woody biomass in order to reduce catastrophic wildland fires and restore forest, woodland, and rangeland health.

In my conservation district, the Nevada Tahoe Conservation District in Stateline, Nevada, forest conditions in areas surrounding Lake Tahoe are indicative of many areas in the Western U.S. experiencing an accumulation of excess fuels leading to reduced resistance to wildfire, disease and insect infestations. These large quantities of biomass are not merchantable as wood products or through other manufacturing industries. However, utilization of this biomass for energy offers a potential economic use for this material, which would help reduce fuel loads.

We recently completed a woody biomass resource and technology assessment for the Lake Tahoe Basin. The study quantifies the Basin's biomass resources and costs, analyzes biomass energy technology performance characteristics, assesses local opportunities for using the material, and summarizes the results of initial planning on a pilot project conducted in conjunction with the Lake Tahoe Unified School District. The study showed there are opportunities for small-scale biomass energy systems to be deployed in the Lake Tahoe Basin.

As a result of the Biomass Feasibility Assessment, sponsored by my District, the Lake Tahoe Unified School District is pursuing funding to purchase a co-generation boiler system to be deployed in the local high school. Biomass to run the new system will supplied by Basin land management agencies from fuels management projects. I have attached a copy of the executive summary of the assessment to my written statement.

In your congressional district, Mr. Chairman, the Deschutes Soil and Water Conservation District received an $89,000 National Fire Plan Community Assistance grant in 2001 to implement an innovative project that turns woody biomass into compost. The grant enabled the district to implement the composting project for Sun River Utilities, which serves Sun River Lodge and Resort and about 4,000 homes in the Sun River development, near Deschutes National Forest. The landowner group was concerned about wildfire and undertook fuels reduction efforts in the lodge pole and ponderosa pine forests. That produced woody biomass from ladder fuels. The organic compost is then sold or spread on Sun River golf courses, building soil composition in an area with volcanic soils that lack organic matter, making it a valued soil additive.

The district continues to focus on initiatives that turn woody biomass "liabilities" into "assets." The district says it needs incentives and marketing capacity to demonstrate value in that material, process it and move it out.

Conservation districts believe efforts such as those I just described and other innovative projects offer tremendous opportunities to reduce catastrophic wildland fires and restore forest, woodland, and rangeland health. In fact, NACD recently entered into a cooperative agreement with the Bureau of Land Management and Forest Service to develop, promote, and improve woody biomass utilization.

Other partners in this effort include the Interior Department's Bureau of Indian Affairs, National Park Service, Fish and Wildlife Service, the cooperative National Fire Plan and the National Association of Resource Conservation & Development Councils.

Under this agreement, NACD is providing resource materials and information to local conservation districts to educate landowners and others on the issue. The goal of this initiative is to help increase public understanding of the social, economic, environmental and aesthetic benefits gained by using woody biomass as a means of reducing fuel buildup on public lands.

We believe more cooperative efforts such as this are needed. Involving local communities and landowners is the ideal way to ensure the success of the Healthy Forests Initiative, the National Fire Plan and other efforts in wildland fire management.

Conservation districts also support other collaborative efforts of the Interior and Agriculture Departments in conducting fuel reduction treatments in the urban wildland interface on federal lands that are at risk from wildfire. To maximize their effectiveness, we believe these collaborative fuels hazard reduction efforts should include:

- A landscape scale approach with the support and involvement of local constituents;
- Cross boundary mitigation;
- Coordination of Federal, state and local government priorities, project design and implementation strategies to maximize effectiveness and minimize costs; and
- Project designs that consider restoration of ecosystem structure, native composition and natural fire regimes.

The drought, which is expected to continue unabated for several more years—especially in the West—adds to the wildland fire issue by contributing to insect and disease problems national forests, BLM lands and private woodlands, as well. Not only is the damage costly to timber, but it also adds to the fuel load.

The nation's conservation districts believe that there are yet many opportunities to develop biomass potential and turn hazardous fuels into useful and valuable products and look forward to continuing our partnerships with the various federal agencies that are responsible for managing the nation's public forests and rangelands.

We appreciate the opportunity to provide the subcommittee with our views.

NOTE: An attachment to Mr. Drew's statement entitled "FINAL REPORT: Biomass Energy Opportunities In and Around the Lake Tahoe Basin" has been retained in the Committee's official files.

————

Mr. WALDEN. Let me start, Lynn, with you. Given the need to treat millions of acres and to maximize the small value-added facilities, how do we find a balance here between large high-volume operators and the small value-added producers? How do we keep a mix? What do you suggest?

Ms. JUNGWIRTH. Well, I think it is relatively simple. You just do it deliberately. You need the mix. But right now, the focus is, of course, on an industrial scale, because everyone keeps telling me because the scale is so hard, large, that we need an industrial solution. Well, ladies and gentlemen, we have no supply. So what you have when you have limited supply is three-megawatt power plants, not 50-megawatt power plants.

Mr. WALDEN. Yes.

Ms. JUNGWIRTH. So let us build on as that supply builds and the large-scale solutions where that is appropriate will come into place. But if you don't deliberately make sure, just like you have an SBA program now for timber, if you don't deliberately make sure that we keep access, if we don't have access, then we are not going to create those local jobs.

So the difference between the company now that we have created in Hayfork that employs 26 people, that is 260 jobs per million board feet. The biomass plant will employ 15 people, but they will burn up 10 million board feet. So you have got to let that mix stay that way. So don't, with your subsidies, encourage something that is going to destroy our competitiveness and our access.

Mr. WALDEN. How do we do that, though? Do you limit the amount somebody gets? Do you put a cap on it?

Ms. JUNGWIRTH. I don't think you need to do that. I think what you need to do is say that the agencies that make that available have to structure their contracts for the whole suite of the industry that is out there, not just the big stuff. And that is the most sustainable way to do it. If you have 30 three-megawatt plants co-located with other wood processors, that is 90 megawatts. But if one of those plants goes offline, you don't care.

Mr. WALDEN. Right.

Ms. JUNGWIRTH. If a 90-megawatt plant goes offline, you are going to care.

Mr. WALDEN. OK.

Ms. JUNGWIRTH. So it can be done deliberately. You have done it before.

Mr. WALDEN. I am intrigued by this notion that somebody mentioned earlier, perhaps on a prior panel, about each region, each forest region having a biomass facility. How do we cause that to occur? Is there a role for the Federal Government in that, because a lot of these regions, the communities—I think of my own district. You have got very remote areas in some cases, very small communities, and yet maybe 50 to 70 percent of the land around these communities is Federal forest or BLM lands that are going to need, clearly need, treatment. Does anybody have any ideas here? I am intrigued by what schools are doing in saving money. I mean, we all know they are pinched. What can we do?

Ms. JUNGWIRTH. Well, you know, the interesting thing about that is if you look at those communities, almost everyone has a vacant mill site—

Mr. WALDEN. Right.

Ms. JUNGWIRTH.—and that means they have power lines going in there.

Mr. WALDEN. Right.

Ms. JUNGWIRTH. They have flat ground.

Mr. WALDEN. Yes.

Ms. JUNGWIRTH. They are on a transportation corridor or they wouldn't be there. The infrastructure, that part of the infrastructure is already there. Frankly, the issue is not do we know what to do with the stuff. We know what to do with this stuff.

Mr. WALDEN. But how do you incent the capital to come in and make that investment, because one of the issues I hear back from people who are in this line of business or would like to be is there is just not the guarantee of supply. And some of these folks were in the mill business and have shut down and auctioned off their equipment after years of retooling down to get to the smallest diameter that is physically possible to make a two-by-two and still there wasn't an adequate supply. Supply is really an underlying problem.

Ms. JUNGWIRTH. And you have to look at the structure of the capital. The three-megawatt plant, you heard him. They expensed it out.

Mr. WALDEN. Right.

Ms. JUNGWIRTH. The bank didn't care.

Mr. WALDEN. All right.

Ms. JUNGWIRTH. You build a 30-megawatt plant, you can't expense it out in 1 day, so that is one of the pieces.

Mr. WALDEN. Mr. Akhtar, did you have a comment you wanted to make?

Mr. AKHTAR. Yes. As I said, my experience is in the technology transfer. On June 17, we organized a symposium at a workshop called "Biorefinery—Value Added Products Out of Wood," and there was an overwhelming response to that. One of the things that we do through the center is sit down with all parties involved, because that is very, very critical because we have quite a few resources that we have to capitalize on, and then put together an effort such as consortium where Federal agencies, State agencies, and the industry—

In the case of Wisconsin, obviously, as I pointed out, is the number one paper producing State in the nation. Now we are trying to work out a deal with the paper industry where we are saying, you can continue making your paper, but at the same time, there is a lot of value that you are losing, so how to capture that. So we are putting together a consortium where we are going to go back to the U.S. Department of Energy for some additional funding to do a demonstration where some of the cost is going to be shared by the paper industry and show a demonstration which could be duplicated to other States, as well.

So it has to be an organization that is dedicated, just like the one we have, that can pull all of these resources together and move forward.

Mr. WALDEN. I see. Will you keep us posted on your progress?

Mr. AKHTAR. Yes, absolutely.

Mr. WALDEN. There are undoubtedly other States that might have that interest or should have that interest.

My 5 minutes is up. Mr. Rehberg?

Mr. REHBERG. Yes. Mr. Akhtar, just out of curiosity, I see in your disclosure requirement you are involved with business that does biopulping. What is that, real quickly?

Mr. AKHTAR. This is a good example. I think one of the Subcommittee members asked this question about technology transfer. This is a new technology that we developed through this consortium effort. Actually at the Forest Products Lab, we started a consortium back in 1987. The technology requires take wood and grow a natural microorganism in about 2 weeks. The fungus secretes enzymes, make the wood soft. So when you make paper, like newspaper or magazine paper, it reduces your energy consumption by about 30 percent and improves the quality.

The other added advantage of that is that it also makes the other chemicals easier to be extracted, and that goes back to the goal of the Committee that we have here. A good example, based on that technology, I personally formed a company called Biopulping International as a spin-off of the technology that Forest Products and the University of Wisconsin developed.

Mr. REHBERG. Thank you. Mr. Coston, could you expand a little bit on your savings? Where was the savings to the school in Darby?

Mr. COSTON. Like someone said, most small communities are beyond the end of natural gas. Darby heated with oil. Their annual heating bill was about $60,000 for the three buildings they have got on their campus. The wood chips—this is a little bit apples and oranges, but the actual cost of their 640 tons of wood chips they burned in the school year just ended was about $18,500. So there is a wide gap between what it costs for wood chips and what it costs for the—

Mr. REHBERG. And that included transportation and there were no changes in the structure of the building itself, so if you were to compare apples and oranges, you could make it work without having to go through what one of the prior panels went through, was determining what kind of a subsidy on biomass, the 1.8 cent per kilowatt.

Mr. COSTON. When we first got into this, we deliberately made all our projections based on private land, some on Montana land, State land. We left the Federal, although most of the land around us is Federal, also. In order to be on the safe side because of the fact that everybody is concerned about it being tied up, we assured ourselves that there was a plentiful supply coming off private land.

Mr. REHBERG. And how long have you determined you have got that plentiful supply?

Mr. COSTON. Well, that 640 tons represents the thinnings off of about 50 acres and there are thousands and thousands and thousands of acres out there that need thinning. Actually, you know, and I recognize up front it is a drop in the bucket. You have got to, like your power plants, you almost have to have every school, if you are just going to concentrate on schools, operating off of biomass heating systems to make any inroads at all into this, but we feel that it is a start.

We chose schools, I guess, mainly because there is no segment of the public structure out there in our part of the world, at least, that is as hard pressed financially as the school district. In order to make—you help every taxpayer in the county if you are able to help out the schools. We looked at prisons, but we thought that maybe "Fuels for Felons" wouldn't be quite as appealing.

[Laughter.]

Mr. REHBERG. Has—

Mr. COSTON. But actually, hospitals and prisons, that type of thing, are—

Mr. REHBERG. Kalispell is looking into nursing homes, and I assume that they have talked to you. If they haven't, we should get you all together.

Mr. COSTON. You are up and down a little bit with schools. You turn the thermostat down at night and turn it down on the weekends and you shut it off in May through September. Hospitals and prisons, something like that is—the harder you work one of these boilers, the more efficient it is.

Mr. REHBERG. Thank you, Mr. Coston, for that.

Mr. WALDEN. And Mr. Renzi waives on questions.

Ladies and gentlemen, thank you very much for your testimony today. It has been quite intriguing and we look forward to working down the road on biomass and see what else we can do to be of assistance, both to our forests and to this emerging and new technology that is out there.

Thank you very much. The record will stay open for additional comments and questions by Members.

I want to insert into the record two statements that have been submitted. The first is from Sherry Barrow of Sherry Barrow Strategies in Ruidoso, New Mexico.

[The prepared statement of Ms. Barrow follows:]

Statement of Sherry Barrow, Sherry Barrow Strategies, Ruidoso, New Mexico

The following is my effort to give you an overview of our business goals and objectives, our progress to date and current constraints with regard to Sherry Barrow Strategies (SBS Wood Shavings) Management & Access to Supply of small diameter timber in Southeastern New Mexico.

The Cree and Scott Abel fires of 2000, the Trap & Skeet Fire of 2001, the Kokopelli, 5/2 and Penasco fires of 2002, a number of Western "burners" and, most recently, the 60,000-plus acre Peppin Fire have brought the reality of catastrophic wild fire to the forefront of regional public awareness.

At SBS, we are interested in the "wholeness of the land"—that is, all the land's values, including timber and other natural resources, wildlife habitat, watershed impacts, recreational opportunity, aesthetics and the results of proper ecosystem management. Sherry Barrow Strategies (SBS) is committed to rural economic development through the creation of a successful small diameter utilization model in

Southeastern New Mexico. SBS is introducing new and effective practices, technologies, equipment and training in order to tackle existing transportation constraints to achieve sustained rural economic development and successful small diameter tree utilization.

SBS business goals are to:

1) Produce wood shavings bedding (SBS Wood Shavings) using small-diameter trees from forest and watershed restoration efforts, utilizing byproducts to co-generate thermal/electrical energy used in the process 2) Identify developing and emerging markets for wood waste products 3) Market, produce, and sell identified value-added products and byproducts to sustain regional economic development 4) Empower community partners in the establishment of sustainable rural economic development by providing access to successful wood waste utilization and value-added biomass models.

In addition, Sherry Barrow Strategies wood utilization business is:

- Assisting in the mitigation of fire hazard by providing a destination point for some of the small diameter trees resulting from treatment of forest and watershed restoration efforts,
- Creating an ecologically sound restoration by-product that will be distributed from Glencoe, New Mexico, expanding to new and emerging markets in future growth phases,
- Reducing the burn time on pile and burn projects, thereby improving regional air quality,
  Did you know: "that for every ton of shavings we put in our bags, we save 3600 pounds of CO2 from going up into the atmosphere.
- Removing insect and disease infested small diameter trees to SBS where the infestations will be heated and destroyed in the processing system, thereby eliminating future damage to healthy stands,
- Bringing to bear an innovative industry that will compliment and, in some cases, help support existing area businesses,
  Did you know:—The January 2003 report prepared by the USDA Forest Service Inventory & Monitoring Institute for the New Mexico EMNRD titled: The Southwest Region's forest-based Community Economic Development Grant Program: Economic Effects in the Apache-Sitgreaves and Lincoln Working Circles, identified that our Lincoln Working Circle "turned" forest industry dollars over 7 to 8 times. The initial economic impact is significant and the secondary impact is as well. We buy services and supplies like: tires, hydraulic fluid, fuel, welding supplies, and services from machine shops, trucking services, and so on.
- Offering technical assistance and training to employees and contractors through a variety of environmental, ecological, and industrial training sessions,
- Serving as a resource for the application of hands-on differentiated curricula for schools in conjunction with local land management agencies, animal husbandry, FFA, YCC and other youth groups, wildlife and watershed groups and other community organizations,
- Empowering community partners though the SBS Outreach Coordinator and the Ruidoso Wild Land Urban Interface Group (RWUIG) in the establishment of a sustainable community effort by providing access to successful wood utilization and value-added biomass models,
- Including media exposure on local wood utilization successes, semi-annual reports to regional municipal, county and economic development councils and, in an effort to heighten public awareness, established a speaker's bureau well-versed in forest and watershed restoration and wood utilization topics.

Federal Funding History

Total Grant funds $547,250

Sherry Barrow Strategies is an active member of the Ruidoso Wild Land Urban Interface Group (RWUIG). RWUIG is a collaborative problem-solving body (LNF, Mescalero Apache Tribe, BIA, BLM, Lincoln County, NM State Land Office, Ruidoso Downs, NM State Forestry, Ruidoso, wood utilization businesses, community groups and other interested entities) empowered to address the health, safety, welfare and economic security of communities at risk of wild fire in the urban interface while respecting the natural interdependence of our ecosystem. Sherry Barrow has served on a Community Forest Management Task Force formed to create fuel reduction ordinances on private property within the Village of Ruidoso. The resulting fire safe guidelines and ordinances from the process have set precedent for community planning in the wild land urban interface and intermix across the west.

SBS Progress to date:

Sherry Barrow Strategies (SBS): Federal/state funding sources, along with personal capital have produced an innovative shavings manufacturing facility utilizing round wood derived from forest and watershed restoration efforts in the geographic region encompassing the Lincoln National Forest in Southeastern New Mexico. The SBS facility is leased from Lincoln County. The processing plant was built within a nine-month period. Commercial production of SBS Wood Shavings began in January of 2003.

Currently, SBS regularly ships semi-truckloads of high quality bagged animal bedding to wholesale/retail locations to multiple states. SBS has a plant labor force of six employees and anticipates adding two more employees this year. We have contracted workers cutting small diameter trees in the forest and 2 truck drivers transporting to SBS year-round. SBS has been working closely with Sierra Contracting, Inc. (SCI), our local composting operation, over the past several months to address transportation constraints for small diameter round wood. SBS is currently paying SCI to transport small diameter trees from treatment sites to SBS Wood Shavings' wood yard in Glencoe, NM. SCI has been operating for several years; has proven ability for "adaptive management"; and has recognized the strength gained from working collectively with other community partners to meet common goals. Once our product is made, SBS also contracts with trucking companies (primarily New Mexico based companies) to transport finished product to wholesale/retail locations in multiple states.

At this time, SBS is using an estimated 337,500 pounds (75 cords) of green round wood per week or 17,550,000 pounds (3900 cords) per year—with the potential to increase usage in the future. SBS estimates that acquisition of 3900 cords will require 1000-1300 accessible acres per year. SBS has utilized green small diameter material from the following sources: LNF-Smokey Bear Ranger District, NM State Trust Land-Moon Mtn., Private landowners—largely projects funded by the WUI dollars through NM-EMNRD Forestry Division, and Municipal Lands-Village of Ruidoso, and the Village of Ruidoso Downs.

Renewable Energy—Co-Generation of thermal heat and electricity:

At Sherry Barrow Strategies we are supportive of these technologies when appropriate economy of scale is observed. We choose to incorporate both thermal heat and electricity generated from wood at SBS Wood Shavings. First,

Thermal: The innovative shaving process at SBS Wood Shavings includes a 12,000,000 Btu sawdust fired burner utilizing the sawdust created in processing to co-generate thermal heat. That thermal heat is then used to dry the wood shavings product before packaging. The burner/dryer system was funded, in part, by a grant (2001) from the Collaborative Forest Restoration Program.

Electricity: Sherry Barrow Strategies restoration wood processing facility in Glencoe (formerly the Glencoe Rural Events Center and Joe Skeen Arena) was ranked first of six locations chosen nation-wide to participate in a Small-Scale Modular Biomass Power System demonstration project utilizing gasification of wood chips, co-sponsored by the U.S. Department of Energy (DOE) through the National Renewable Energy Lab (NREL) in Littleton, CO., Community Power Corporation, Golden, CO., and the USDA Forest Service, Forest Products Laboratory (FPL), Madison, WI. The unit was rolled out in late 2002.

The small, modular biomass unit processes wood chips from fuel reduction projects creating electricity and thermal heat for the SBS facility in Glencoe, NM. If you have questions about the program or the reasons for our #1 ranking, you may contact Sue LeVan-Green at the Forest Products Laboratory—Program Mgr., S&PF Technology Marketing Unit. Her contact information is: slevan@fs.fed.us or you may phone her at (608) 231-9518.

As for the economic impact of grants to forest based industry, please see the January 2003 report prepared by the USDA Forest Service Inventory & Monitoring Institute for the New Mexico EMNRD titled: The Southwest Region's forest-based Community Economic Development Grant Program: Economic Effects in the Apache-Sitgreaves and Lincoln Working Circles.

LOCAL SUPPLY/ACCESS TO SMALL DIAMETER WOOD

Due to the threat from catastrophic wild fire in the urban interface and intermix, the USDA Forest Service-LNF has identified a need for thinning one-third of the 200,000 acres in the Sacramento Ranger District and 70,000 acres in the Smokey Bear Ranger District. Forest Service figures show the Lincoln National Forest (LNF) growing an average of 30 to 40 million board feet per year with a loss on average of 7 million board feet to insects. These figures do not include the potential for loss from fire and other catastrophic events. (reference: Dennis Watson, Timber Management Officer, LNF). In accordance with current funding plans, LNF estimates 2500 to 3500 acres per year will be made available for pre-commercial thinning. Restoration wood from small diameter treatments will be made available for wood utilization. (reference: Brian Power, Aviation and Fire Officer—LNF). In light of the Healthy Forests Initiative, SBS expects some modification of these plans may occur.

New Mexico State Forestry—Capitan District has received National Fire Plan WUI funds for fuel reduction treatment (small diameter) on private lands. The Capitan District Forester has identified approximately 1500 acres for fuels reduction treatments in priority areas within the wild land urban interface and the work is now under way.

The Village of Ruidoso—The Village has implemented a low-intensity thinning project in the Grindstone Lake recreation area. In the summer of 2002, the Village of Ruidoso began a 438 acre restoration project adjacent to the 3000 acre LNF—Smokey Bear Ranger District "Eagle Creek" project. The "Eagle Creek" project has received federal funding from Collaborative Forest Restoration Program. On the Village's 438 acre project, an estimated 60 yards per acre of woody biomass (under 5" dbh) and approximately 3 cords per acre of round wood (5" to 12" dbh) were slated for removal over a two year period.

The Village of Ruidoso Downs—The Village will begin restoration of eighty acres in the Village watershed area this year. Sherry Barrow Strategies will be removing round wood to SBS Wood Shavings in Glencoe, NM, for utilization.

Additional projects are pending in conjunction with: New Mexico State Trust Lands and the Bureau of Land Management.

Resources

For us, the value of the Forest Products Laboratory (FPL), in Madison Wisconsin, the Southern Research Station (SRS) in Auburn Alabama, have been beyond measure. The Marketing and Technology resource provided by FPL and the equipment and systems research from the SRS have been an essential element of our innovation and success. Those of us working toward solutions in reducing the threat of catastrophic wildfire by building service capacity and rural economic development through wood utilization businesses rely on the expertise and resources provided by both Labs.

The research component provided by Sue LeVan-Green (FPL) and Robert Rummer (SRS) has absolutely saved Sherry Barrow Strategies at least two years in mistakes and money. Without the research provided, we would have had to do months of "trial and error" research and testing. The staff is responsive and has performed beyond our expectations. I understand that Rural Development Economic Action Program (EAP) funds have been an integral support of the Forest Products Lab and the Southern Research Station and the other Research Stations. I am concerned that the future of these vital resources may be in jeopardy without the restoration of EAP funding.

The Roswell office of the Small Business Development Center—Gene Simmons, Director, has assisted with business planning over the last three years.

With regard to NM-Forestry Division Four Corners Sustainable Forests Partnership (FCSFP): The Partnership, which U.S. Senator Pete Domenici is given credit for fostering, quickly became our "clearinghouse" for growth and development resources and mentoring. It is important to note that without exception, FCSFP has been the only program with an integrated plan from the "Stump-to-the-Consumer". Early on, the FCSFP Program Manager provided a flow-chart which helped us to understand funding streams, the timeframes for the paperwork, and the economies of scale in the forest industry. Resources were shared across boundaries (like the Forest Products Lab and Southern Research Station), as well as entrepreneurial resources and marketing expertise on a national and international scale. This program has focused on the impact of sustainable forest-based communities and fostered "working circles" of interdependent small businesses with new and emerging markets for round wood.

Under the auspices of EMNRD-Forestry Division, the Four Corners Sustainable Forests Partnership, has provided countless hours of resource information, contacts, problem solving, federal funding sources, access to mentors and encouragement through the Partnership. The Partnership is evolving this year into the Southwest Sustainable Forests Partnership, targeting the needs of Arizona and New Mexico.

The Collaborative Forest Restoration Program, propelled largely by U.S. Senator Jeff Bingaman, has also focused on the need for diverse collaborative stakeholders. A strong focus on environmental impacts, appropriate fire regimes, and preservation of old and large trees has garnered a new awareness of the "wholeness" of the land and the long-term effects of a more balanced approach to restoration practices.

The Technical Advisory Panel deliberation process is open to the public. Observing the deliberation process is a valuable educational experience. Program Manager, Walter Dunn has provided a rare opportunity for potential grantees to learn about diverse perspectives on forest restoration. The panelists have become resource conduits for our work. We now have a number of "go-to" resource people in different areas across the country. Our involvement with CFRP convinced us that we bear a responsibility for the treatment side of the small diameter trees we utilize for products.

While the CFRP does not have the strong market-side focus that exists with the Southwest Sustainable Forests Partnership, its collaborative environmental strength on the treatment-side begets valuable assets for our resulting wood products in the market.

CONSTRAINTS

In order to facilitate sustainable rural economic development, forest health, and complete the "stump to consumer" cycle, community partners must have tools to build infrastructure and successful systems. A collaborative effort toward building service capacity, including technical assistance and training for environmentally sensitive equipment and appropriate small diameter handling systems is the next step toward long-term sustainability. The Lincoln National Forest has demonstrated a willingness to explore all available contracting options including Stewardship contracts in order to meet management objectives. Long-range access to forest biomass is the next step toward long-term sustainability.

Recent federal funding has planted the seeds for emerging small diameter wood businesses. SBS believes our community will establish sustainable forest-based businesses suitable for replication in other western states.

This work is not for the faint at heart. We are building a foundation for long-term sustainable forest management. No one entity can do it alone. We need to have all the stakeholders involved. In the beginning, collaborative community groups were guarded in attempting to form relationships—some fell apart and regrouped and others just backed away from what they believed was destined to fail. First, we had to build tolerance, then establish a dialogue, and identify common ground and then work collectively within our "zone of agreement". So, it takes time.

The SBS experience with FS, BLM, BIA, New Mexico State Trust, and NM Forestry Division staff has been extremely positive and we are making solid progress toward our goals. In the LNF region, we also have the ever-present threat of wild fire. Our entire community acknowledges the danger and we are working together toward forest and watershed restoration.

TRANSPORTATION

Currently SBS is moving away from handling small diameter trees too many times with inappropriate equipment and systems. The results are encouraging. Still, transportation cost of the trees from the prescribed treatment site to a utilization site remains a regional constraint. We had hoped the transportation $20 per green ton credit in the Energy Bill would give some interim relief. If available, it would have doubled the transport range for small diameter wood. SBS is rare in that we are a regional small diameter processing facility with an established, stable, year-round outlet for green small diameter timber.

With regard to access to supply of small diameter trees, I see several promising opportunities: Better management practices, more effective contracting instruments, new low-impact cost effective forestry equipment, equipment capable of accessing areas previously deemed inaccessible in our region, and a heightened public awareness resulting in strong support for fuel reduction in the WUI and watersheds.

In addition to traditional products, the use of biomass and other waste as a renewable energy is long overdue. There are plans for building everything from 5kw to 35megawatt power plants to wood chip-retrofitted community boiler systems.

We must address the need in rural communities for economic diversity and appropriate scale. As for biomass power plants, SBS believes that 1/2 to 1 megawatt plants strategically located near the wood supply and an end-user seem more reasonable.

While we believe in sustainable communities, we are concerned that the desire to reduce forest fuel loading could result in a push for a "quick-fix" solution. I do not want to see small business diversity left out of the "mix" by the creation of an over-scaled biomass facility. Nor do I want unnecessary tree cutting to feed a business under the "guise" of restoration. Huge power plants are expensive to build and expensive to maintain. Infrastructure to deliver power is expensive, can be invasive, and, finally, who will buy the power? And, will the power be purchased at a rate that will pay for the investment?

When faced with a choice today—and we are using both thermal and electric heat generated from wood at our facility—I see thermal heat generation as less risky to communities and less expensive to incorporate into existing infrastructure.

Again, I urge caution and vigilant attention to the selection of appropriately scaled endeavors. Whatever solutions are realized, an environmentally sensitive, diverse economy driven by healthy forests is Sherry Barrow Strategies answer for sustainable rural communities.

Thank you again for your diligence. I hope you find this information of interest. I will be pleased to take any questions you may have.

———

Mr. WALDEN. I would also like to insert a statement submitted by Todd Brinkmeyer, President and Owner of Plummer Forest Products, Inc., in Plummer, Idaho.

[The prepared statement of Mr. Brinkmeyer follows:]

Statement of Todd Brinkmeyer, President and Owner, Plummer Forest Products, Inc., Plummer, Idaho

Chairman Walden and members of the subcommittee.

I am Todd Brinkmeyer, President and owner of Plummer Forest Products, Inc. located in Plummer, Idaho. Plummer is a small community of 900 residents in the heart of the Coeur d'Alene Indian Reservation in the panhandle of Idaho, approximately 35 miles south of the city of Coeur d'Alene, Idaho.

I started Plummer Forest Products in 2000 on the site of a former large log sawmill vacated by Rayonier Company in 1998.

Plummer Forest Products is a fully integrated biomass to energy facility and small log sawmill producing 5 megawatts of electricity and eighty million board feet of lumber per year. We have 85 employees. After struggling for three years of start-up challenges, I can you proudly say that—for now "Plummer Forest Products is a viable enterprise.

I appreciate the opportunity to testify before the subcommittee today.

As the written submission for the record, I am including a summary of a presentation I have made at the National Bio Energy and Wood Products Conference sponsored by the U.S. Departments of Interior, Agriculture and Energy in Denver, Colorado on January 21, 2004. That presentation provides detail on the history, current configuration and challenges faced by my company as we have sought a way to profitable convert small logs and trees into wood products and energy.

In short, Biomass to energy works, but not as a stand-alone enterprise. The sawmill and energy plant must work together. The cost structure associated with removing woody biomass from the forest, hauling the material to a facility and converting the fiber into a product suitable for electricity production is prohibitive without massive subsidization.

Plummer Forest Products has developed a program that includes removing the woody biomass from the forest by leaving the material attached to a small log segment that can be separated at our facility and further processed into stud lumber and other building materials. By leaving the material attached, the handling cost and freight can be reduced to a level that makes the integrated process viable.

Thank you for your time, and I hope that we can take this model and others like it, refine them and develop a prescription for federal lands that reduces wild fire risk, promotes healthy forests, jobs in rural communities, and does not cost the tax payers money. All the pieces are in place to do that in some area

[NOTE: An attachment to Mr. Brinkmeyer's statement has been retained in the Committee's official files.]

———

Mr. WALDEN. If there is no further business to come before the Subcommittee I again want to thank the members of the Subcommittee who participated today and we stand adjourned.

[Whereupon, at 3:44 p.m., the Subcommittee was adjourned.]

[Additional information submitted for the record follows:]

Statement of Dr. Liam E. Leightley, [1] **Department of Forest Products, Forest and Wildlife Research Center, Mississippi State University**

WHY THE SOUTH EAST UNITED STATES REGION NEEDS TO PURSUE THE PRODUCTION OF ENERGY AND CHEMICALS FROM WOOD BIOMASS

Forest Biomass energy has been considered as an industry of the future, providing potential new markets for forest thinnings, residues and waste (Quick, 2003). The potential for this industry is continuing to increase because changes in the Global economy have caused a reduction in demand for timber in the United States, especially in the southeast where the demand for pulpwood has significantly decreased. This has led to local industry experiencing high inventories and lower prices for small diameter pinewood and thousands of acres of overstocked pine plantations. A biomass energy industry could utilize wood considered to be un-merchantable or underutilized and could contribute to alleviating the nation's economic, energy and environmental concerns. The large inventory of small trees could be reduced, stumpage prices could be increased and the value of forest assets restored. The removal of such material from the forests, as pre-commercial thinnings would also create healthier forests that were less susceptible to attack by destructive insects and disease. In addition to the biomass obtained from thinnings, a significant volume of solid wood waste is produced by the wood products industry each year. This wood waste could also be used as a feedstock source of biomass for energy production.

A recent article appeared in the South Carolina Forestry Association Journal (SFCA, June 2004) stating that without new uses and markets for our trees, there is little incentive to continue to invest in growing timber. The article referred to the fact that a Bio-based Industry Alliance was formed May 21, 2004 in Tuscaloosa, AL, to capitalize on the resources and strengths of the Southeast United States region to provide a stimulus to the rural economy, reduce our dependence on fossil fuels and to develop the science of extracting chemicals from forest and farm crops. The feedstock for the new energy and chemicals industry will be low-value timber, forest and farm residues and farm crops. The need for the Alliance was summarized by Gene Quick, an organizer of the Alliance and is presented here in its entirety. The summary clearly states the need for biomass utilization by a number of interests, not the least of which are small to medium size landowners. It is those interests which could derive direct benefit from the utilization of wood biomass.

"For over 70 years we planted trees for what we thought was a growing and never-ending demand. Timber prices were strong and in a reliable, upward trend, with only an occasional pause. For many years we exported chips to meet the demands of foreign markets. In the last few years things have changed dramatically. Some pulp mills have closed and others have reduced production. In some markets we can import pulp cheaper than we can produce it domestically. It appears unlikely that another pulp mill will be built in this country. And now large volumes of wood chips are being imported through the port of Mobile, AL, further reducing demand for—and prices of—our own small diameter timber. Without new uses and markets for our trees, there is little incentive to continue to invest in growing timber. Farming has been on the decline even longer. The creation of a bio-based energy and chemicals industry creates new opportunities for all those involved in all phases of the growing and harvesting of farm and forest products.

The University of Alabama's Alabama Institute for Manufacturing Excellence (AIME) is the home of this new regional Alliance of industry, university, federal and state government agencies, private businesses, forest and farm landowners, and landowner associations across the southeastern United States.

There are 29 organizations and companies, from South Carolina, Georgia, Alabama, Mississippi, Louisiana, and Texas, now participating in the Alliance.

While there are differences in economic circumstances among states, the rural economies in all thirteen of the Southeastern United States have suffered from the closure and slowed production rates of pulp mills and the decrease in value of farm crops. Concurrently, the need for energy and the dependence on foreign oil continue to increase, driving up the use and cost of energy from fossil fuels.

The Southeast, with its 214 million acres of forestland and 128 million of the nation's 338 million acres of total farmland, has renewable, expandable, and sustainable sources of energy and chemical feedstocks. The region also has an under-utilized labor force, the business infrastructure, and the scientific resources needed to bring the concept of a bio-based industry into reality.

With the wise use of this enormous land resource, combined with our technical and business capabilities, it will be possible to significantly reduce our dependence

---

[1] Chairman, Southern Alliance for the Utilization of Biomass Resources

on fossil fuels without degrading air and water quality or compromising our timber and food supplies, while also creating jobs in our rural economy.

The bio-based energy and chemicals industry will create a high-volume, non-cyclical market for forest and farm crops. When well established, it will revive the depressed timber market and create a demand for agricultural crops.

The Alliance brings together the resources, researchers, government agencies, and business interests necessary to make rural development in the Southeastern U.S. a reality. Creation of this industry will reduce our dependence on fossil fuels, much of which is imported, by using forest and farm products which are renewable, sustainable, and expandable. The potential positive impacts on the economy and environment are substantial. The Alliance will be the conduit for collaboration, co-ordination, communications and actions, which will result in bringing much needed change".

A roadmap for Agriculture Biomass Feedstock Supply in the United States was recently published in a report by the U.S. Department of Energy Efficiency and Renewable Energy, Biomass Program (DOE, 2003). The report stated a goal that biomass will supply 5% of the nation's power, 20% of transportation fuels, and 25% of chemicals by 2030. A key concept on which the roadmap is based upon is the—Biorefinery—which processes biomass into value added product streams. The roadmap considered that it will be necessary for USDA Laboratories and the Nation's Universities to develop the science and technology base for the biorefinery as well as address the important knowledge gaps that have been identified. The benefits of using biomass to drive a biorefinery supplying domestically produced power, fuels and products were considered to be significant, including decreased demand for imported oil, revenue to a depressed agriculture industry and revitalized rural economies. Four high level goals were identified for the feedstock required for the biorefinery concept—Biomass Availability, Sustainability, Feedstock Infrastructure and System Profitability. Currently, the primary biomass resource is obtained from wood waste produced by forest products industries. The amount of this resource could be significantly increased by using thinnings material removed from forests for fire hazard reduction in the Western states and for improving the health of the forests in Southern states.

Conversion of Wood Biomass into Energy and Products

Wood biomass can be converted into a range of products using several different processes. The biomass can be burnt to produce energy or heat, converted into fuels which can be burnt to produce heat or power or used to produce chemicals and materials. There are a large number of technologies under development which could become commercial and provide the needed conversion routes for the wood biomass.

The South's Industrial Forest Products Biomass

Changes in global trade in wood products have resulted in reduced demand for wood pulp produced in the U.S. Numerous U.S. pulp and paper mills have been closed in recent years and industry experts predict no new capital investment for U.S. mills due to environmental concerns and international competition. Sawmills that previously depended on revenue from the sale of chips for pulp have seen chip prices dramatically decline in recent years. A survey of 12 sawmills located in the southeastern U.S. shows that chip prices have declined from about \$24/ton in 1990 to \$21.50/ton in 2002 (Rountree, 2003). This 11-percent price decline becomes an approximate 20-percent decrease if normal inflation is considered.

Solid wood waste has been classified as municipal solid waste, construction and demolition debris, primary timber processing mill residues and logging residues (McKeever, 2003). The National volume of waste wood generated was estimated to be some 230 Million dry tons, with 125 Million tons being combusted and not used and 104 Million tons available for recovery. Of the 104 Million recoverable tons 52 Million tons was generated in the South.

Utilization of industrial sawdust and bark for energy biomass has been practiced by industrial forest products companies for centuries. Typically, sawmill lumber dry kilns utilize the steam generated by burning less desirable wood waste, mainly bark and sawdust. Higher-value wood chips from green lumber edgings and trimmings have had much higher value as feedstock for the production of pulp and paper than for energy. However, lower demand for industrial wood chips as pulp and paper operations have been reduced has resulted in lower prices for industrial wood chips. For this reason utilization of industrial wood chips as well as bark and sawdust for energy production may now be feasible.

The South's Plantation Pine Resource

Plantation pine silvical practices have been adopted for a rapidly increasing share of timberlands in the South as shown by the increased annual acreage of trees planted over time in Figure 1. From 1952 to 1996, 57 million acres of pines were planted in the southern U.S. (Smith et al. 2000). In 1994, total privately owned plantation pine acreage in the South was 30 million acres or about 47 percent (Siry and Bailey 2003) of the total forested privately owned acres. Modern plantation pine silvical practices call for planting of genetically improved seedlings that grow 16-percent faster than unimproved stock. Wide, between-seedling spacing followed by early thinning has resulted in rapidly increased growth rates (Zobel and Jett 1995). Siry and Bailey estimate that rate of pine growth in the south increased an average of 2.6 percent annually between 1987 to 1994.

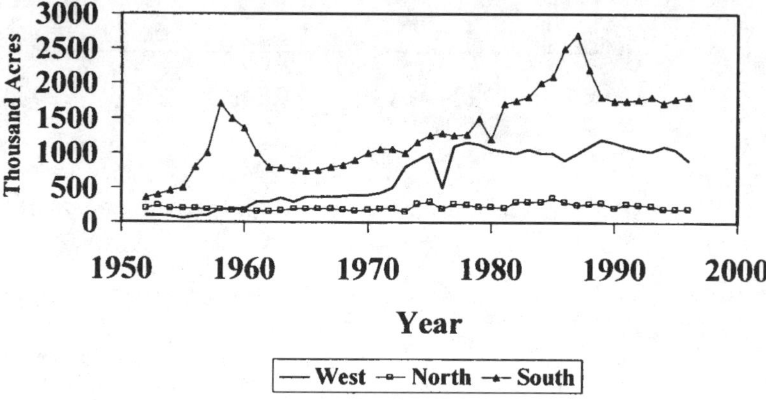

Figure 1.   Annual forest land planted annually by region of the U. S. from 1950 to 1996. (Source: Smith et al. 2001).

Thinning is a forest management practice which removes small trees to reduce stand density and improve the quality of merchantable stems. Fast-grown plantation pine thinnings frequently contain up to 80 percent of their volume in juvenile wood (Zobel and Sprague 1998). Presence of a large percentage volume of juvenile wood in young southern pine stems results in serious problems in utilization of the material harvested. Juvenile wood is characterized by lower density, lower transverse shrinkage, higher longitudinal shrinkage, lower strength, thinner latewood bands, more compression wood, higher initial moisture content, thinner cell walls and lower cellulose to lignin ratio (Bendtsen 1978). Pulp yields from juvenile wood are lower and lumber is considerably weaker and very prone to warp (Zobel and Sprague 1998).

Juvenile wood is contained in approximately the first 10 growth rings of pine tree stems. As plantation pine trees add mature wood, following this initial 10-year period, the relative percentage of juvenile wood decreases such that utilization problems from older trees are reduced. For this reason, the most severe and objectionable utilization problems occur in trees from first and second thinnings rather than older sawlog-sized timber.

The described utilization problems for fast-grown plantation pine are particularly severe for wood from first thinnings which typically contain a very high proportion of juvenile wood. A survey of Mississippi's wood industry found that many companies are restricting purchase of timber to ages above 17 years because of the high percentage of juvenile wood contained in younger timber (Stiglbauer, P. 2002). This restriction has resulted in landowners encountering difficulty in having their timber thinned in accordance with their planned harvest schedule.

Pulpwood stumpage prices have declined even more dramatically than industrial wood chip prices as a result of slackening demand for pulp feedstock. Figure 2 shows that prices declined from over $10 per green ton in 1997 to $6.50 per green ton in 2002 (Rountree 2003). If inflation is factored into these prices the value of pine pulpwood stumpage has declined by about 50 percent in 5 years. Siry and Bailey (2003) predict that increased supply and slack demand will result in low southern pine pulpwood stumpage prices through the year 2030.

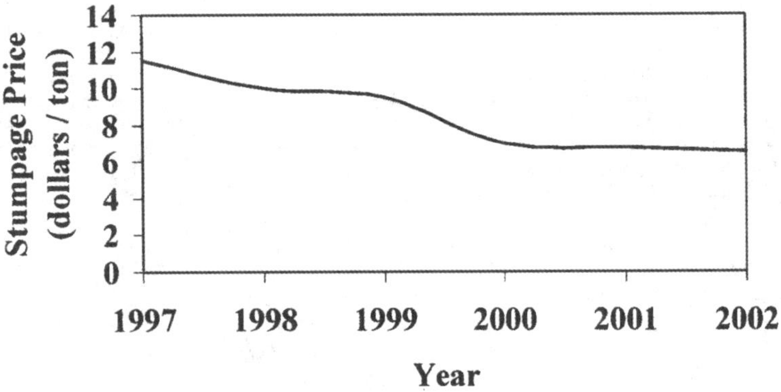

Figure 2. Pulpwood stumpage price trends over the preceding 5 years.

Forest Energy Plantations

Considerable research has been performed to develop short rotation intensive culture (SRIC) forestry plantations for energy. Traditionally, this research has focused on production of energy from fast-growing hardwood species such as eastern cottonwood, American sycamore, sweetgum, willow and non-native species such as the eucalypts (Bruce 1994). Until recent years the value of pine plantation thinnings for pulp and paper feedstock has been so high that utilization of this resource for energy has been prohibitive. However, current and future economic trends indicate that utilization of pine thinnings for energy feedstock is becoming a viable alternative.

Lack of perceived economic viability has limited the research performed for utilization of pine plantation materials for biomass production. Eight-to-10 year rotations for plantations are typically applied when managing hardwood stands for biomass (Portland 1994). If thinned by a similar early harvest schedule the harvest of plantation pine at age 10 for biomass would release residual pine stems to increase their growth rates with the rate increase roughly proportional to the severity of thinning. Faster growth after 10 years of age would act to solve the juvenile wood problem by increasing the percentage of mature wood in relation to the juvenile wood core. By contrast, pine plantation first thinning removal for pulpwood is usually practiced on stands at about 15 years of age. In addition to the earlier increased growth of the residual stand there are increased economic benefits to landowners if income from thinnings occurs earlier in the rotation (Bullard and Straka 1998).

Based on the silviculture applied to produce, it is probable that whole-tree chipping will also be the most practical and economical harvesting method for pines. Largest volume of biomass and the least amount of handling of stems would occur if needles, branches, bark and stems are harvested and utilized for a value-added product. As Table 1 indicates, the inclusion of needles and top wood components will increase the moisture of the biomass to some degree. Branches will have no influence on moisture content and inclusion of bark will reduce the total biomass moisture content substantially. Net biomass moisture content will be a variable function of the volume that each component represents for each tree. However, a net moisture content of about 125 percent may represent a practical working average.

Table 1. Moisture content of slash pine tree components (Source: Koch 1992).

Component	**Moisture Content**
Needles	154
Top	153
Stem	116
Branches	115
Bark	67

Biomass and Energy Availability

The volume of thinning material available from the south's timberlands can only be inferred from the available data. Forest inventory data do not provide volume information on trees less than 5.0-inch diameter breast height (dbh). The volume of the dbh class most likely to be utilized for energy is the 5.0 to 6.9-inch class which is the minimum size for which data are available. The volume of material contained in this dbh class is 11.4 billion ft3 (Smith et al. 2001). A conservative assumption is that the available volume of material from smaller diameters of 2 to 4.9-inch dbh stems is equal to that contained in the 11.4 billion ft3 value for the 5.0 to 6.9-inch dbh class. This results in an estimate of available biomass from all pine stands of about 23 billion ft3. The volume available in plantations is approximately 47 percent of this value (Siry and Bailey 2003), or10.8 billion ft3. If 20 percent of total plantation volume is thinned, the total biomass currently available for removal from application of these systems is 2.2 billion ft2.

While considerable research has been performed to determine silvical and harvest volumes for hardwood species, only test sites have resulted. No large-scale hardwood biomass energy plantations are available. However, development of viable fuel markets would result in application of the research performed over the three decades.

The net usable heat from combustion of one pound of dry wood is 4300 Btu. Green wood at 100-percent moisture content provides slightly more than 70 percent of this value at 3020 Btu (Koch 1992).

BioOil can be taken as an example of a potential liquid fuel obtained from wood biomass. The percentage yield of BioOil varies with the process applied but ranges from 40 to 75 percent with 60 percent agreed on by most practitioners as a safe estimate for systems designed to maximize BioOil yield.

The heating value per pound of BioOil is 6800 Btu/lb (Bridgewater et al. 1999). At 20-percent moisture content wood weighs 35.9 lbs/ft3 which results in 21.5 lbs of BioOil per ft3 of wood to give 146,200 Btu of energy. Therefore, the 2.2 billion ft3 of pine available would provide 3.2 x 1014 Btus of energy if converted to BioOil.

Acknowledgements.

I would like to thank Mr. Gene Quick, Forest Energy Associates and Dr Phillip Steele, Department of forest products, Mississippi State University, for providing me with information and data appearing in this testimony.

References

Bendtsen, B. A. 1978. Properties of wood from improved and intensively managed trees. Forest Prod. J. 28 (10): 61-72.

Bridgewater, A., C. Czernik, J. Diebold, D. Meir, P. Radlein. 1999. Fast Pyrolysis of Biomass: a Handbook. CPL Scientific Publishing Services, Ltd. Newbury, UK. 188 p.

Bruce, A. P. 1994. Short rotation forestry in Loblolly pines. In Proc. Of the Short Rotation Forestry in Loblolly pines. March 1-3, Mobile, AL.

Bullard, S.H. and T.J. Straka. 1998. Basic Concepts in Forest Valuation and Investment. Preuda Education and Training. Auburn, AL. 270 p.

Roadmap for Agriculture BioMass Feedstock supply in the United States. DOE, November 2003.

Koch, P. 1992. Utilization of the southern pines, Vol. II: Processing. USDA Forest Service, Southern Forest Experiment Station. U.S. Govt. Printing Office, Washington, DC.

McKeever, D. 2003. Taking Inventory of Woody Residuals. Biocycle. July 2003, 31—35

Portland, C. J. 1994. Utilization of cottonwood plantations. In Proc. Of the Mechanization in Short Rotation Intensive Culture Forestry Conference. March 1-3. Mobile, AL.

Quick, G. 2004. Alliance formed to Advance Wood Biomass Energy. South Carolina Forestry Association Journal 24(5) and (6).

Rountree, S. 2003. Chip prices at member sawmills. Unpublished report. Southeastern Lumber Manufacturers Association. Forest Park, GA. 1 p.

Siry, J. P. and R. L. Bailey. 2003. Increasing southern pine growth and its implications for regional wood supply. Forest Prod. J. 53(1): 32-37.

Smith, W. B., J. L. Vissage, D. R. Darr and R. Sheffield. 2001. Forest Resources of the United States. USDA Forest Service, North Central Research Station, St. Paul, MN. 190 p.

Stiglbauer, P. 2002. The status and utilization of plantation pine timber in Mississippi. M. S. Thesis. Department of Forest Products, Mississippi State University. 68 p.

Zobel, B. J. and J. B. Jett. 1995. Genetics of Wood Production. Springer-Verlag, New York. 337 p.

Zobel, B. J. and J. R. Sprague. 1998. Juvenile wood in forest trees. Springer Publications. New York. 300 p.

[A letter submitted for the record by The Honorable Janet Napolitano, Governor, State of Arizona, and The Honorable Dirk Kempthorne, Governor, State of Idaho, on behalf of the Western Governors' Association, follows:]

WESTERN GOVERNORS' ASSOCIATION

Bill Owens
Governor of Colorado
Chair

Janet Napolitano
Governor of Arizona
Vice Chair

Pam O. Inmann
Executive Director

Headquarters:
1515 Cleveland Place
Suite 200
Denver, Colorado 80202-5114

303-623-9378
Fax 303-534-7309

Washington, D.C. Office:
400 N. Capitol Street, N.W.
Suite 388
Washington, D.C. 20001

202-624-5402
Fax 202-624-7707
www.westgov.org

July 8, 2004

The Honorable Greg Walden
Chairman
House Resources Subcommittee on Forests and Forest Health
1337 Longworth House Office Building
Washington, DC   20515-6205

The Honorable Jay Inslee
Ranking Member
House Resources Subcommittee on Forests and Forest Health
1337 Longworth House Office Building
Washington, DC   20515-6205

Dear Congressmen Walden and Inslee:

On behalf of the Western Governors' Association we would like to thank the Subcommittee for taking the initiative to hold a hearing on June 23rd on biomass utilization. We welcome the timely discussion by the Subcommittee on this issue as Western Governors believe that biomass utilization could be a contributor to restoring our forests to a healthy condition with the added benefit of helping to diversify our energy portfolio through a renewable energy source.

Recent wildfires throughout the West demonstrate the current unhealthy state of our forests and wildlands. Hazardous fuel loads are at all-time highs. Combined with ongoing drought and the wildland-urban interface where people and fuels meet and mix, we have a volatile situation and a strong need to implement fuels reduction projects to restore our forests to a healthy condition. Biomass utilization may be a part of the solution. The 10-year Comprehensive Wildfire Strategy and Implementation Plan, supported by the WGA, highlights the importance of hazardous fuel reduction and biomass utilization. Finding uses for materials resulting from hazardous fuels treatments could help provide economic development opportunities for rural communities while creating safer and healthier communities and watersheds. We must underscore that the management goal for our forests must be the long-term health of our forests, but we firmly believe biomass can be a constructive part of achieving that goal.

The Honorable Greg Walden
The Honorable Jay Inslee
July 8, 2004
Page 2

The WGA is directly involved with the application of biomass technologies through the National Biomass State and Regional Partnership established through the Department of Energy. The WGA, along with the other regional bodies in the partnership, will work toward developing increased bio-energy awareness, increased bio-energy development, improved leveraging of investments, and state policy and incentives favorable to biomass. As one mechanism, we think this biomass partnership has great promise as the linkage between reducing forest health threats and finding clean, secure and diversified energy sources.

The Governors thank you for your efforts and for holding a hearing on this important topic.

Sincerely,

Janet Napolitano
Governor of Arizona

Dirk Kempthorne
Governor of Idaho

cc: Rep. Pombo
        Rep. Rahall
        Western Governors
        The Hon. Joshua Bolten
        The Hon. Ann Veneman
        The Hon. Gale Norton
        The Hon. Spencer Abraham